Breeding of Horticultural Crops
Principles and Practices

The Author's

Dr. Bijendra Kumar Singh is presently working as Assistant Professor, Department of Horticulture, Tilak Dhari Post Graduate College, Jaunpur (U.P.). He did his Graduation in Agriculture from VBSPU, Jaunpur, Post Graduation in Horticulture from BBAU, Lucknow and Ph.D. in Horticulture from BHU, Varanasi. He has specialization in Pomology.

Dr. Singh has served Tilak Dhari Post Graduate College in various capacities like Member of Campus Greenery of college under IQAC committee and Academic Counselor of IGNOU study centre. Dr. Singh has 152 publications (research papers 30, proceeding papers 3, conference/seminar papers 52, books 5, book chapters 25 and popular/semi-scientific articles 37). He had been convener of the organizing committee of One National Workshop. Dr. Singh qualified ICAR NET in Fruit Science and received University Grants Commission (UGC) fellowship during his Ph.D. He is a life member of renowned societies and journals in India like Asian PGPR Society of Sustainable Agriculture, International College of Nutrition's, Society for Advancement of Research on Pomegranate, Trends in Biotechnology & Biological Sciences, Advances in Life Sciences and Trends in Biosciences. He has supervised 4 M.Sc. (Ag.) and 6 Ph.D. students as Chairman and Member of Advisory Committee.

Dr. Rajaneesh Singh is presently working as Associate Professor and Head, Department of Horticulture, Tilak Dhari Post Graduate College, Jaunpur (U.P.). He did his Graduation in Agriculture from Udai Pratap Autonomous College, Varanasi, Post Graduation in Horticulture from G.B. Pant University of Agriculture and Technology, Pantnagar and Ph.D. in Horticulture from BHU, Varanasi.

Dr. Singh has served Tilak Dhari Post Graduate College in various capacities like In-charge Garden Section, Captain in NCC, Member of the Sports Committee, Academic Counselor of IGNOU study center and Former Proctor of the college. Dr. Singh has 19 year teaching experience of horticulture. Dr. Singh has more than 71 publications to his credit (research papers 25, proceeding papers 6, conference/seminar papers 12, books 5, book chapters 10 and popular/semi-scientific articles 13). He had been chairman of the organizing committee of One National Workshop. Dr. Singh qualified ICAR NET in Horticulture and Vegetable Science and received Senior Research Fellowship in ICFRE during Ph.D. programme. He is fellows of Indian Society of Vegetable Science and Indian Society of Horticulture Science. He has supervised 4 M.Sc. (Ag.) and 5 Ph.D. students as Chairman of Advisory Committee.

Breeding of Horticultural Crops
Principles and Practices

Bijendra Kumar Singh
Rajaneesh Singh

2021
Daya Publishing House®
A Division of
Astral International Pvt. Ltd.
New Delhi – 110 002

ISBN : 9789390371808 (Int Edition)

Published by : **Daya Publishing House®**
A Division of
Astral International Pvt. Ltd.
– ISO 9001:2015 Certified Company –
4736/23, Ansari Road, Darya Ganj,
New Delhi-110 002
Ph. 011-43549197, 23278134
E-mail: info@astralint.com
Website: www.astralint.com

World Noni Research Foundation

64, Third Cross Street, Second Main Road, Gandhi Nagar, Adyar, Chennai – 600 020

Telefax: 044-2442 3601 Email: mail@worldnoni.org Website: www.worldnoni.org

Dr. Kirti Singh

FNASc, FNAAS, FNABS

CHAIRPERSON

Foreword

There are many challenges in production of horticultural crops in meeting the daily requirement of fruits and vegetables for growing population in the limited area, reduction in cultivable land, poor yield potential and susceptibility to insect and pest and diseases. Breeding of horticultural crops is a complete and comprehensive resource for the development of new cultivars or clones of horticultural crops. Specific traditional breeding methods are also covered, with an emphasis on how these methods are adapted for horticultural species. In addition, the integration of biotechnology with traditional breeding methodology has been explored, with an emphasis on specific application for fruits, vegetables and ornamental crop species.

The present book entitled ***"Breeding of Horticultural Crops: Principles and Practices"*** deals with basic elementary information which is essential at undergraduate and postgraduate levels for the students of horticulture. This book fulfills the gap and provides all aspects of the fundamentals of horticultural crop breeding and seed production technology of vegetable crops.

In my opinion, this text book shall be extremely helpful to the students of horticulture as well as to the teachers of horticulture. I have great pleasure in congratulating the authors for their enthusiasm and hard work in bringing out extremely useful and informative publication.

Date: *February 2020*

Place: *Jaunpur*

(Kirti Singh)

Former Vice-Chancellor

Preface

The pressure of an ever increasing population and periodic famine due to unexpected flood and drought has forced and awakened the horticultural scientist evolve new plant types for diversified use. Besides some limitations in the improvement of horticultural crops such as long juvenile phase, high heterozygosity, limited information on inheritance pattern, parthenocarpy, and lesser number of seed per fruit, hybridization, selection, mutation, and other tools of horticulture breeding have resulted in the development of a number of varieties in fruits and vegetable crops for various purpose.

The present book entitled **Breeding of Horticultural Crops- Principles and Practices,** Section I is ventured with the objective to provide latest possible information's on fundamentals of breeding of horticultural crop including Centre of origin, Mode of reproduction, Breeding system, Breeding methods, Breeding for Pest and Disease Resistance, and Intellectual property rights. Section II including Cytotaxonomy and Breeding Behaviour of Fruit Crops, *i.e.,* Mango, guava, citrus, papaya, pineapple, pomegranate, and temperate fruits. Section III refers to Breeding Behaviour of Vegetable Crops, *i.e.,* Tomato, brinjal, chilli, cole crops, okra, pea, and cucurbit crops. Section IV includes Vegetable Seed Production Technology, *i.e.,* Seed Production Techniques, Development of Seeds, Seed Production of Vegetable Crops, and Hybrid Seed Production.

We thank Dr. Kirti Singh, Former Vice-Chancellor, for the inspiration given by him in writing this book. Authors are highly grateful to Prof. Anil K. Singh (BHU), Prof. Ranveer Singh, Prof. J.P. Tiwari, Prof. Hari Har Ram, Dr. Hari Baksh, and Mr. Raj Pandey for their encouragement and keen interest.

Dr Bijendra Kumar Singh
Dr Rajaneesh Singh

Contents

Section-II
Cytotaxonomy and Breeding Behaviour of Fruit Crops

Section-III
Breeding Behaviour of Vegetable Crops

Section-I

Fundamentals of Horticultural Crop Breeding

Chapter 1

Introduction

Plant Breeding

It can be defined as an art and science and technology of improving the genetic makeup of plants in relation to their economic use for the mankind. In other word plant breeding is branch of agriculture that focuses on manipulation plant heredity to develop new and improved plant types for use by society. People in society are aware and appreciate the enormous diversity in plants and plant products. They have preferences for certain varieties of agricultural and horticultural crops.

Aim of Plant Breeding

It aims to improve the characteristic of plants so that they become more desirable agronomically and economically. The specific objectives may vary greatly depending on the crops under consideration.

Scope of Plant Breeding

The plant breeding has been helping the mankind with knowledge of classical genetics, number of varieties have been evolved in different crop plants. Since the population is increasing at an alarming rate there is need to strengthen the food production which is serious challenge to these scientists concerned with the agriculture or horticulture. Advances in molecular biology have sharped the tools of the breeders and brightened the prospects of confidence to serve the humanity. The application of biotechnology to field crop has already led to the field testing and commercialization of genetically modified crop plant.

Scientist Contribution

Name of Scientist	*Contributions*
Hull	Recurrent selection for SCA
Jenkins	Recurrent selection for GCA
Davenport	Dominance hypothesis
Flor	Gene for gene hypothesis
Johansen	Pure line concept
Harlan and Pope	Back cross method
Jones and Davis	Cytoplasmic genetic made sterility
Levee's	Self-incompatibility
Vavilov	Centre of diversity
Vilmorin	Progeny test
Hugo de Vries	Mutation
Nilsson and Ehle	Bulk method
Shull	Heterosis
Jones and Davis	Male sterility

Indian

B.P. Pal, M.S. Swaminathan, H.K. Jain, N.G.P. Rao, B.R. Murti, E.A. Siddiq, V.L. Chopra, V. Arunachalam, Vishnu Swarup and Puskarnath.

Horticultural Crop Breeding

"It deals with genetic improvement of various horticultural crops such as fruits, vegetables and ornamental plants. In other word horticultural crop breeding is the science of the production of new crop varieties which are superior to their parents in horticultural crops. Superior crop varieties can be evolved by means of selection, introduction, hybridization, polyploidy, mutation, tissue culture and biotechnology."

Vegetable Breeding

"Vegetable breeding is the art and science of changing and improving the heredity of vegetable plants. In other word vegetable breeding consist of principles and methods required for favourably changing and improving the genetic constitution of plants in relation to their economic use".

Fruit Breeding

"Fruit breeding is the manipulation of a biological system in fruit crops that requires many generations to achieve results. It is also dynamic exciting and challenging profession, operating under continually changing conditions."

Improvement of fruit crops is difficult because a long gestation period, high heterozygosity, and scanty information on inheritance pattern, often cross-pollination and less number of seeds restricting availability of hybrid seedlings for evaluation. In future search for desired gene, critical study of inheritance pattern and use of biotechnological tools require due attention in combining ideal characteristics in varietal improvement programme of fruit crops.

Major Problems of Fruit Breeding

1. It has long generation cycle of 2-10 years depending upon species and cultivars.
2. Fruit crop has long juvenile period due to that early assessment of strains are not possible, *e.g.,* Mango, Jamun, *etc.*
3. Majority of the fruit species are highly heterozygous therefore large population is required for selection.
4. Most of fruit species are polyploid in nature, *e.g.,* Ber, Banana, *etc.*
5. Polyembroyony is quite common, *e.g.,* Mango, Citrus, *etc.*
6. Several fruit crop are parthenocarpic resulting into seedlessness, *e.g.,* Banana, Pineapple, *etc.*
7. Abundance of sexual incompatibility, *e.g.,* Mango, Apple, Pear, Loquat, *etc.*
8. More number of chromosome, creating a big problem in genetical analysis, *e.g.,* Ber, Mulberry, *etc.*
9. Presence of single seed in single fruit, requiring more number of crossing, *e.g.,* Mango, Litchi, Mahua, *etc.*

Objective of Plant Breeding

The main objective of breeding is to get maximum quality production per unit area with low cost and free from biotic and abiotic stresses.

- ☆ **Higher yield:** The plant breeder aims to improve the yield of economic produce.
- ☆ **Improved quality:** The price of produce is determined by its quality again quality differs from crop to crop.
- ☆ **Biotic resistance:** Crop plants are attacked by various diseases and insect pests resulting in considerable yield losses. Genetic resistance is the cheapest and the best method of minimizing such losses.
- ☆ **Abiotic resistance:** Crop plants also suffer from abiotic factors such as drought, soil salinity, heat, wind, cold and frost. Breeder has to develop resistant varieties for such environmental conditions.
- ☆ **Earliness:** It is most desirable characters which has several advantages. It requires less crop management. Facilitates double cropping system and reduces overall production cost.

- ☆ **Synchronous maturity:** It refers crop maturity at one time.
- ☆ **Removal of toxic compounds:** It is essential to develop varieties free from toxic compound in some crops to make them safe for human.
- ☆ **Wider adaptability:** Adaptability refers to suitability of a variety for general cultivation over a wide range of environmental conditions.
- ☆ **Photo insensitivity:** It is also found in vegetable crops.
- ☆ **Extended harvest period:** The harvesting of fruit and vegetable are extended for longer usage.

Objective of Fruit Breeding

Its main objective is to study of best quality fruit production for longer usase and freedom from the biotic and abiotic stresses.

(A) For Root Stock

1. Wide geographical adaptability.
2. Easily propagation through asexual reproduction.
3. It should be resistant to biotic and abiotic stresses.
4. It should be free from suckering habit.
5. Plant should not have brittle root system, *e.g.*, EM-9 (Rootstock Apple).

(B) For Scion Cultivars

1. Dwarf growth stature.
2. High productivity with good quality.
3. Attractive fruit colour with pleasant aroma.
4. Suitable for processing and export.

India is bestowed with a wide range of agro-climatic and soil conditions. Therefore almost all types of fruits can be grown in one or the other part of the country. India is second largest producer of fruits and vegetables next to China. In India horticultural crops occupy about 7.2 per cent of gross area, contribute about 18-20 per cent gross value of agricultural output and 5 per cent of export earning in agriculture. More focus on search for desired genes, critical study of inheritance pattern and use of biotechnological tools are needed in combining ideal characteristics in varietal improvement programme of fruit crops.

History of Horticultural Research

Horticultural research in India was started at the departments of botany in six agricultural colleges established in 1905 at Pune, Coimbatore, Lyallpur, Nagpur, Sabour and Kanpur. Almost at the same time the Imperial Agricultural Research Institute was set up at Pusa (Bihar) and the provincial and central departments of agriculture were organized which were to look after the work on horticultural crops. At that time the responsibility of research on horticultural crops was mainly of the state governments. During this period some of European settlers like Lee in Kullu Valley, Coutts and Stokes in Shimla hills and some European Missionaries in South India introduced several cultivars of fruit crops from UK, France and East India, *etc.* A pomological station was established of Coonoor near Ooty in 1920 to study the adaptability of temperate fruit varieties and other vegetable crops.

Importance of Horticultural Crop Breeding

In view of the importance of horticultural crops in daily diet and the need to increase their supply position to growing population, more emphasis was laid on horticulture research during the sixties. Especially during the last 15 years development in horticulture has gradually moved from rural to urban areas and from traditional agricultural enterprise to corporate sector adapting improved technology, greater commercialization and professionalism in the management of production and marketing. Intensive research in horticulture has been taken up in many ICAR institutions, agricultural universities for the last 50 years with the result that many improved cultivars have been made available for planting by the horticulturist despite the fact that the problems encountered in breeding of horticultural crops are enormous.

1. Social function.
2. General utility and nutritional value.
3. Economical importance.
4. Trade and industrial value.
5. Ecological balance and pollution control.

Chapter 2

Centres of Origin

Introduction

N.I. Vavilov (1951), a Russian Scientist surveyed the world and divided the whole universe into eight main centres of origin. Origin in simple words means how a particular fruit and vegetable has been originated and domesticated for the welfare of living being. A centre of origin is the place where maximum dominant factors exist in the population of the horticultural crops and maximum diversity of that particular vegetable. The recessive genes are on periphery of those centres of origin. These were the main convection followed the recessive genes which arise by mutations in nature. This is also called primary centre of origin. When two or more species crossed with each other and artificial natural selection occurs in the population, it is called secondary centre of origin, for example, several species of *Brassica* and *Allium*. Harlan (1954) called these small pockets or valley's small local areas where maximum diversity plus dominant genes existed as micro-centers.

Primary centre of origin: Areas where cultivated plant species are supposed to have originated. Centre of origin is based upon centre of diversity.

Secondary centre of origin: In some areas, certain crop species show considerable diversity but they had not originated there, such areas are known as secondary centre of origin.

Centre of diversity: Areas where cultivated plant species show much greater variation than any other place of the world.

Gene centre	Primary centre	Secondary centre
Chinese-Japanese	Brinjal, Wax gourd, Cabbage	Watermelon, Amaranths
Indo-Chinese	Swiss gourd, Gourd, Bean, Cucumber	Cabbage, Bottle gourd
Hindustan centre	Brinjal, Cucumber, Guard, Drumstick, Okra	Watermelon, Bottle gaurd
Central Asia	Onion, Garlic, Carrot, Spinach	Brinjal, Watermelon, Cauliflower, Okra
Near East	Onion, Garlic, Beet	Okra
Mediterranean	Cabbage, Cauliflower, Radish	Sweet Pepper, Garlic, Okra
African	Brinjal, Melon, Cowpea,	Onion, Lima bean, Okra
Central America and Mexican Region	Tomato, Pepper, Sweet Potato	
South America	Tomato, Cassava	Common Bean
Region	Pumpkin	

The concept of centres of origin was conceived by Nikolai Ivanovich Vavilov (N.I. Vavilov) based on his studies of a vast collection of plants at the Institute of Plant Industry, Leningrad during his tenure as director from 1916-1936. According to Vavilov, crop plants evolved from wild species in the areas showing great diversity and termed them as primary centres of origin. But in some areas, certain crop species show considerable diversity of forms although, they have not originated from such areas which are known as secondary centres of origin. Eight main centres of origin are recognized as proposed by Vavilov.

1. **China:** This is the one of the largest and oldest centres of origin. It includes mountainous parts of Central and Western China beside, neighbouring low lands, *e.g.,* Pear, Peach, Apricot, Plum, Mandarin, Litchi, Tea, *etc.*
2. **Hindustan:** This centre includes Burma, Assam, Malayan, Archipelago, Java, Sumatra and Philippines. But this centre does not include North West India, Punjab and North Western Frontier Provinces. Later on this centres of origin is divided into Indo-burma and Siam Malaya Java centres of origin, *e.g.,* Mango, Sour lime, Mandarin, Coconut, Banana, Jackfruit, Bael, Jamum, Aonla, Karonda, *etc.*
3. **Central Asia/Afghanistan:** It includes North West India (Punjab North Frontier Provinces and Kashmir) All Afghanistan, Soviet Republics of Tajikistan and Uzbekistan, *e.g.,* Grape, Almond, Apricot, Pear, Pistachio nut, Apple, *etc.*
4. **Asia Minor/Caucasus/Trans-caucasus:** It includes the interior of Asia Minor the whole of Transcaucasia, Iran and high land Turkmenistan. The centre is also known as the Near East or Persian centres of origin.
   - ✰ **Primary centre-** Fig, Pomegranate, Apple, Grape, *etc.*
   - ✰ **Secondary centre-** Chestnut, Pistachio nut, Walnut, *etc.*

5. **Mediterranean centre:** Peppermint.
6. **Abyssinian:** It includes Ethiopia and hilly country of Eritrea, *e.g.,* Coffee.
7. **Central American:** This includes region of South Mexico and Central America. It is also known as Mexican centres of origin, *e.g.,* Guava, Papaya, Rubber, Avocado, *etc.*
8. **South America:** This centre includes the high mount regions of Peru, Bolivia, Ecuador, Colombia, part of Chile and Brazil and whole of Paraguay. Further this centre was sub-divided into three centers, *i.e.,* Peru, Chile and Brazil Paraguay centres of origin. For example, Pineapple and Cashew are from Chile and Brazil, Guava is from Peru and Chile. It is also called as Tropical America.

Regions in India with Rich Horticultural Genetic Resources Diversity

The National Bureau of Plant Genetic Resources 1976 (NBPGR) New Delhi is the nodal institute working on survey collection, exchange, quarantine, characterization, evaluation, conservation and documentation of PGR including HGR. It plays a pivotal role in crop improvement, development and diversification of agriculture in India through germplasm introduction from various foreign sources collection within the country and abroad and germplasm supply to plant breeders and other users. National Gene Bank (NGB) was established in 1996 with a strong capacity of 1 million seed accessions. Well equipped cryo-preservation and *in-situ* (*in-vitro*) conservation facilities were developed to cater to the conservation of HGR especially recalcitrant seed species and vegetatively propagated materials in 1986. Some of the most important commercial crop cultivated extensively in India today are introduction from other countries. Para rubber was first introduced from Brazil in 1873. Tapioca introduced into India by Portuguese and the Dutch Cinchona introduce into Nilgiris from Peru in 1860. Coffee introduced in to India in 1700 by Muslims who returned from a pilgrimage to Mecca.

Chapter 3

Modes of Reproduction

Introduction

The reproduction system is principally responsible for perpetuation and preservation of a particular genotype. Further the mode of reproduction also determines the genetic constitution of horticultural crop whether it is homozygous or heterozygous.

Knowledge of the mode of reproduction and pollination is essential for a plant breeder, because these aspects help in deciding the breeding procedures to be used for the genetic improvement of a crop species. In other word, choice of breeding procedure depends on the mode of reproduction and pollination of a crop species. This chapter deals with two important aspects, *viz.,* (1) mode of reproduction (2) mode of pollination.

Reproduction refers to the process by which living organisms give rise to the offspring of similar kinds/species. In crop plants, the mode of reproduction is of two types:, *viz.,* (1) sexual reproduction (2) asexual reproduction.

Types of Reproduction in Horticultural Crops

A. Asexual Reproduction

Multiplication of plants without the fusion of male and female gametes is known as asexual reproduction. Asexual reproduction can occur either by vegetative plant parts or by vegetative embryos which develop without sexual fusion (apomixis). In it mainly vegetative parts are used for the reproduction of the plant. It is of two types: *viz.,* (1) apomixis (2) vegetative reproduction.

1. *Apomixis*

Apomixis refers to the development of seed without sexual fusion (fertilization).

In apomixis embryo develops without fertilization. Thus apomixis is an asexual means of reproduction. Apomixis is found in many crop species. Reproduction in some species occurs only by apomixis. This apomixis is termed as obligate apomixis. But in some species sexual reproduction also occurs in addition to apomixis. Such apomixis is known as facultative apomixis. There are four types of apomixis:, *viz.*, (1) parthenogenesis, (2) apogamy, (3) apospory (4) adventive embryony.

- ✰ **Parthenogenesis-** Parthenogenesis refers to development of embryo from the egg cell without fertilization.
- ✰ **Apogamy-** The origin of embryo from either synergids or antipodal cells of the embryo sac is called as apogamy.
- ✰ **Apospory-** In apospory, first diploid cell of ovule lying outside the embryo sac develops into another embryo sac without reduction. The embryo then develops directly from the diploid egg cell without fertilization.
- ✰ **Adventive embryony-** The development of embryo directly from the diploid cells of ovule lying outside the embryo sac belonging to either nucellus or integuments is referred to as adventive embryony.

2. *Reproduction by the Vegetative Parts of Plant*

Vegetative reproduction refers to multiplication of plants by means of various vegetative plant parts. Vegetative reproduction is again of two types:, *viz.*, (i) natural vegetative reproduction (ii) artificial vegetative reproduction.

- ✰ **Natural vegetative reproduction:** In nature, multiplication of certain plants occurs by underground stems, sub-aerial stems, roots and bulbils. In some crop species, underground stems (a modified group of stems) give rise to new plants. Underground stems are of four types:, *viz.*, rhizome, tuber, corm and bulb. The examples of plants which reproduce by means of underground stems are given below:

  - ❒ **Rhizome-** Turmeric, Ginger, Colocasia, *etc.*
  - ❒ **Tuber-** Potato.
  - ❒ **Corm-** Arvi, Bunda, Gladiolus, *etc.*
  - ❒ **Bulb-** Garlic, Onion, *etc.*

  Sub-aerial stems include runner, sucker, stolen, *etc.* These stems lead to vegetative reproduction in mint, rose, strawberry (runner), pineapple, banana (sucker), *etc.* Bulbils are modified forms of flower, they develop into plants when fall on the ground. Bulbils are found in garlic.

- ✰ **Artificial vegetative reproduction:** Multiplication of plants by vegetative parts through artificial method is known as artificial vegetative reproduction. Such reproduction occurs by cuttings of stem and roots, and by layering and grafting. Examples of such reproduction are given below:

(a) Cutting

Cutting is piece of plant which produces adventitious root (in case of stem and leaf cutting) and a new shoot system (in case of root cutting) when produced in rooting media and keeping all other condition as favourable. It gives rise to a new plant almost true to mother plant.

(i). Stem Cutting

- **Hard wood cutting:** Hard wood cutting has diameter 2 cm and each cutting has 2-3 nodes. All the cutting performs well in misting of water and exogenous application of auxin, *e.g.,* Grape, Fig, Mulbery, Guava, Mango, Quince, Currant and Pomegranate.
- **Semi hard wood cutting:** Prepared from partialy mature but tender woody shoot. Wood cutting is taken from lignified nature woody shoot. Length may vary from 20-25 cm, *e.g.,* Citrus and Olives.
- **Soft wood cutting:** Soft wood cuttings are those prepared from soft succulent terminal portion of woody perennial. It is desirable to retain some leaves for photosynthesis but more leaves are undesirable which may cause moisture loss due to transpiration. Soft wood cuttings are 8-15 cm long with 2-3 nodes. In deciduous plant cutting should be taken in March-April and in evergreen plant, cutting should be taken in May-Sep. Rooting in cutting starts within 15-30 days if temperature is 20-25°C and relative humidity 80 per cent, *e.g.,* Crab apple, Apple, Peach, Pear, Plum, Apricot and Cherry.
- **Herbaceous cutting:** Herbaceous cuttings are tender, succulent and leafy part of stem of herbaceous plant. The terminal 7-12 cm of healthy shoot is cut and basal leaves are removed and upper leaves remain undisturbed, *e.g.,* most of the ornamental plants.

(ii). Root Cutting

For example Apple, Crab apple, Sweet potato, Citrus, Lemon, *etc.* layering and grafting are used in fruit and ornamental crops.

(iii). Leaf Cutting

For example *Begonia, Bryophyllum, Sansevieria, etc.*

(b) Layering

Layering involves the regeneration of adventitious root from a shoot when it is still attached with the mother plant. Exogenus application of auxin has been found to hasten the initiation of root, increase percentage of rooting and number of roots per layers. This is more common in guava, litchi, citrus, *etc.*

(*i*). **Air layering**, *e.g.,* Litchi and Loquat.

(*ii*). **Stool layering**, *e.g.,* Guava, Apple rootstock and Pear.

(*iii*). **Trench layering**, *e.g.,* Apple and Pear rootstock.

(iv). Compound layering, *e.g.*, Muscadine Grape.

(v). Tip layering, *e.g.*, Blackberry, Raspberry and Gooseberry.

(c) *Grafting and Budding*

Mango, Guava, Chestnut, Cashewnut, Aonla, Bael, Ber, Citrus, Grape, Apple, Pear, Plum, Almond, *etc.*

B. *Sexual Reproduction*

It is also known as amphimixis in which multiplication is done through seed that develop normal meiosis followed by syngamy (fertilization), *e.g.*, Phalsa, Karonda and Papaya. Multiplication of plants through embryos which have developed by fusion of male and female gametes is known as sexual reproduction. All the seed propagating species belong to this group.

- ☆ **Sporogenesis:** Production of microspores and megaspores is known as sporogenesis. In anthers, microspores are formed through microsporogensis and in ovules the megaspores are formed through megasporogenesis.
- ☆ **Microsporogenesis:** The sporophytic cells in the pollen sacs of anther which undergo meiotic division to form haploid, *i.e.*, microspores are called microspore mother cell (MMC) or pollen mother cell (PMC) and the process is called microsporogenesis. Each PMC produce four microspores and each microspore after thickening of the wall transforms into pollen grain.
- ☆ **Microgametogenesis:** This is nothing but the production of male gametes or sperm. On maturation of the pollen, the microspore nucleus divides mitotically to produce a generative and a vegetative or tube nucleus. The pollen is generally released in this binucleate stage. The reach of pollen over the stigma is called pollination. After the pollination, the pollen germinates. The pollen tube enters the stigma and travels down the style. The generative nucleus at this phase undergoes another mitotic division to produce two male gametes or sperm nuclei. The pollen along with the pollen tube possessing a pair of sperm nuclei is called microgametophyte. The pollen tube enters the embryo sac through micropyle and discharges the two sperm nuclei.
- ☆ **Gametogenesis:** The production of male and female gametes in the microspores and megaspores is known as gametogenesis.
- ☆ **Megasporogensis:** A single sporophytic cell inside the ovule, which undergo meiotic division to form haploid megaspore, is called megaspore mother cell (MMC) and the process is called megasporogenesis. Each MMC produces four megaspores out of which three degenerate resulting in a single functional megaspore.

- **Megagametogensis:** The nuclear cell of a functional megaspore divides mitotically to produce four or more nuclei in which three mitotic divisions produce eight nuclei. Three of these nuclei move to pole and produce central egg cell and two synergids cell. Megaspore develops into a mature megagametophyte or embryo sac and the development of embryo sac from a megaspore is known as megagametogensis.
- **Fertlization (syngamy):** Fusion of one of two sperms with egg cell producing a diploid zygote is known as fertilization. In other word the fusion of nuclei from the pollen grain with nuclei in the ovule. Fertilization allows the flower to develop seeds. In other words fertilization is the fusion of male gamete with female gamete. The fusion of the remaining sperm with the secondary nucleus leading to the formation of a triploid primary endosperm nucleus is termed as triple fusion. The primary endosperm nucleus after several mitotic divisions develops into mature endosperm, which nourishes the developing embryo.
- **Post fertilization development:** A series of changes in the ovule follows as fertilization is over. Synergids and antipodal cell become disorganized and the egg cell is covered with a cell wall known as Oospores. This follows developed seeds.

Floral Biology and Development of Gametophytes

When androecium (male) and gynoecium (female) both are present in same flower, this is called as hermaphrodite or bisexual, *e.g.,* ber, citrus, guava and sapota. Some flower posses either the male or female reproductive parts it is known as unisexual flower. When male and female parts of flowers are present in separate flower but on same plant it is known as monoecious (unisexual) condition, *e.g.,* aonla, walnut, hazelnut and jackfruit. But when they are present on different plants separately, it is known as dioecious condition, *e.g.,* papaya, date palm and kiwifruit. Andromonoecious condition is found in mango where male and hermaphrodite flower are present in same panicle. Further the gynodioecious condition is found in some cultivars of papaya where some hermaphrodite flowers are found on plant bearing pistillate flower (*Pusa delicious* and *Pusa majestic*) and fig.

Flower

The outstanding and most significant feature of the flowering plant is the flower. Flower is the highly specialized made up by stem and leaves.

Kinds of Flower

- **Complete flower:** Contains all floral organs, *i.e.,* sepals, petals, stamens and pistil, *e.g.,* tomato, chillies, cole crops, brinjal, root crops, *etc.*
- **Incomplete flower:** Lack of one or more of these floral organs, *e.g.,* cucurbits, asparagus, *etc.*

☆ **Perfect flower:** Consists of both stamens and pistil in the same flower, *e.g.*, tomato, chillies, cole crops, brinjal, root crops, *etc.*

☆ **Imperfect flower:** Lack of either male organs or female organs, *e.g.*, cucurbits, asparagus, papaya, date palm, *etc.*

The imperfect flowers are two types:

1. **Monoecious:** When male and female flowers are separate but present in the same plant, for example, caster.
2. **Dioecious:** When staminate and pistilate flowers are present on different plants, then it is called dioecy, *e.g.*, papaya, date palm, *etc.*

The amount of natural crossing or natural selfing within these crops may vary with (i) the variety, (ii) seasonal condition, (iii) The direction and velocity of the wind, and (iv) the insect population. Thus the vegetable breeder working with self and cross pollinated crops will need to determine the extent of the natural selfing/ crossing under particular condition.

Flower Parts

There are following parts in a flower:

1. **Peduncle:** The floral stalk, stem supporting the flower, sometimes referred to as the pedicel.
2. **Receptacle:** Modified floral stem or axis from which arise the floral appendages or modified leaves.
3. **Sepal:** It is also called as calyx. The outer most whorl of leaves, typically green and protect the inner floral parts in buds collectively.
4. **Petal:** It is also called as corolla. The second whorl of leaves typically brightly coloured and assists in attracting pollinators collectively.
5. **Perianth:** Collective term of sepal and petals if perianth part cannot be differentiated into sepals and petals, that is they look so much alike then they are called tepals.
6. **Filament:** Slender stalk of the stamen supporting the anther, permits exertion of pollen out of flower.
7. **Anther:** Fertile portion of stamen that dehisces to release pollen grains, composed of anther sacs.
8. **Stamen:** The male structure of flower comprising filament and anther, collectively all the stamens are referred to as the androecium (house of male).
9. **Ovary:** Basal portion of pistil that contains ovules at maturity become fruit with seeds.
10. **Ovules:** Fertile portions of pistil that contain female gametophyte (embryo sac) and develop into seeds after fertilizations.

11. **Pistil:** Flask-shaped female structure often referred to as carpels, all pistils (one or more) are referred to as gynoecium (house of female).
    - ☆ **Androecium-** The male reproductive organ of the flower is called androecium where stamen is unit of androecium (anther and filament) found in male flowers.
    - ☆ **Gynoecium-** The female reproductive organ of the flower, consisting of one or more carpels is called gynoecium. Carpels are consist of stigma, style, ovary.

Flower Bud Differentiation

It is also known as flower induction for bud differentiation to occur. There is enhanced cell division in the central zone below the apical part of the meristem. Due to cell division there is differentiation of parenchyma cells which surround the meristem into flower primordial, *e.g.*, Mango (October), Apple (June-July) and Pear (July- Augast).

- ☆ **Simple Bud-** It grows into vegetative shoot. It is also known as leaf bud.
- ☆ **Mixed Bud-** It grows into shoot with a flower. It is also known as flower bud.
- ☆ **Compound Bud-** It grows into both leaf and flower.
- ☆ **Solitary Flower-** Solitary flower are borne singly and separate from one another, *e.g.*, Saucer Magnolia and Blood Red Geranium (*Geranium sanguineum*).

Inflorescence

In some species of plants many small flower are borne together in dense cluster called inflorescence. These are used for arrangement of flower on shoots, a useful plant ID tool. Many types of inflorescence are found in plants.

1. **Spike:** Individual flowers are sessile (without pedicel), lower flowers open first in indeterminate plants, *e.g.*, Gladiolus "Lavadandy".
2. **Raceme:** Individual flowers have pedicels and pedicels can vary in length from species to species. Lower flowers open first in indeterminate, *e.g.*, Virginia sweet spire and Bleeding heart.
3. **Panicle:** A highly branched inflorescence consisting of many repeating units. Panicle can be made by many spikes, racemes, corymbs or umbels. It is indeterminate, *e.g.*, Common lilac and Ohio Buckeye.
4. **Umbel:** Individual flower pedicels all originate from the same spot on the peduncle, outer flowers open first, for example, *Allium*.
5. **Corymb:** Somewhat similar to umbel. Individual flower pedicels are attached to the peduncle at different point and often flat topped, outer flower are first to open, *e.g.*, Callery Pear.

6. **Cyme:** In real life cymes tend to be flat or convex shaped and inner flowers first open and found determinate., cymes are often compound, *e.g., Judd viburnum* and *Arrouwood viburnum.*
7. **Corymb *vs* cymes:** Often these two types of inflorescence can look a lot alike. Corymbe are indeterminate they continue to elongate as the season progresses, cymes are determinate and do not continue to elongate as the growing season progresses, also the inner flowers of cymes open first.
8. **Composite head:** Consists of separate ray and disk florets, bracts may be green but can also be coloured (straw flower) and after pollination and fertilization each disk flower becomes a single seeded fruit, *e.g.,* Sunflower and Purple coneflower.

The andromonoecious condition are found in mango where male and hermaphrodite flowers are present in same panicles. Further the gynoecium condition is found in some cultivars of fruit crops, *e.g.,* Papaya (*Pusa majestic* and *Pusa delicious*) and Fig.

Mode of Pollination

Transfer of pollen grain from anther to stigma is termed as pollination. In other word, the transfer of pollen from the anther (male parts of flower) to the stigma (female parts of a flower) of the same flower or of another flower is called pollination. Pollination is a prerequisite for fertilization. It is of two types:

1. Self-pollination or Autogamy

It takes place when pollen grain from an anther falls on the stigma of same flower on same plant (variety), *e.g.,* Guava and Citrus. In other word transfer of pollen grains from the anther to the stigma of same flower is known as autogamy or self-pollination. Autogamy is the closest form of inbreeding. Autogamy leads to homozygosity. Such species develop homozygous balance and do not exhibit significant inbreeding depression on selfing.

Mechanism Promoting Self-pollination

- ✰ **Bisexuality:** Presence of male and female organs in the same flower is known as bisexuality. The presence of bisexual flowers is a must for self-pollination. All the self-pollinated plants have hermaphrodite flowers or perfect flower.
- ✰ **Cleistogamy:** When pollination and fertilization occur in unopened flower bud, it is known as cleistogamy. It ensures self-pollination and prevents cross-pollination, *e.g.,* Grape and Sapota, Lettuce, Frech bean and pea.
- ✰ **Homogamy:** Maturation of anthers and stigma of a flower at the same time is called homogamy. As a rule, homogamy is essential for self-pollination, *e.g.,* Citrus, Peach and Apricot, Tomato.

- **Chasmogamy:** Opening of flowers only after the completion of pollination is known as chasmogamy. This also promotes self-pollination and is found in crops like tomato, brinjal, chilli, capsicum, *etc.*
- **Position of anthers:** In some species, stigmas are surrounded by anthers in such a way that self-pollination is ensured. Such situation is found in tomato and brinjal. In some legumes, the stamens and stigma are enclosed by the petals in such a way that self-pollination is ensured, *e.g.,* pea.

Main Features of Self-pollination

- The transfer of pollen grain from anther to stigma of same flower (autogamy) or to the stigma of another flower of same plant (geitonogamy).
- Need no external agency or agent.
- Male and female parts mature at same time (homogamy).
- New varieties cannot be produced, new variations are not possible, less vigorous, offsprings are produced and thus cannot adapt to environmental change, *e.g.,* Bean, Cowpea, Lettuce, Tomato, *etc.*

2. Cross-pollination or Allogamy

In cross-pollinated species maximum 5 per cent self-pollination may take place. If pollen grain from flower of a plant is transferred to the stigma of flower of another plant, it is known as cross-pollination. In other word transfer of pollen grains from the anther of one plant to the stigma of the flower of another plant is called allogamy or cross-pollination. This is the common form of outbreeding. Allogamy leads to heterozygosity. Such species develop heterozygous balance and exhibit significant inbreeding depression on selfing.

Mechanism Promoting Cross-pollination

- **Dicliny:** It refers to unisexual flowers. This is of two types:, *viz.,* (i) monoecy (ii) dioecy. When male and female flowers are separate but present in the same plants, it is known as monoecy. In some crops, the male and female flowers are present in the same inflorescence such as in mango, castor and banana. In some cases, they are on separate inflorescence as in maize. Other examples are cucurbits, grapes, strawberry, cassava and rubber. When staminate and pistillate flowers are present on different plants, it is called dioecy. It includes papaya, date palm, spinach and asparagus.
- **Dichogamy** (from the Greek *dikho*-apart and *gamous*-marriage): It refers to maturation of anthers and stigma of the same flowers at different times. Dichogamy promotes cross-pollination even in the hermaphrodite species. Dichogamy is of two types: *viz.,* (i) protogyny (ii) protandry. When pistil matures before anthers, it is called protogyny such as in banana, pomegranate, plum, *etc.* When anthers mature before pistil, it is known as protandry. It is found in coconut, walnut, sapota, sugarbeet, *etc.*

- **Heterostyly:** When styles and filaments in a flower are of different lengths, it is called heterostyly. It promotes cross-pollination, such as sapota, litchi, pomegranate, brinjal, *etc.*
- **Herkogamy:** Hinderance to self-pollination due to some physical barriers such as presence of hyline membrane around the anther is known as herkogamy. Such membrane does not allow the dehiscence of pollen and prevents self-pollination such as in lima bean.
- **Self-incompatibility:** The inability of fertile pollens to fertilize the same flower is referred to as self-incompatibility. It prevents self-pollination and promotes cross-pollination. Self-incompatibility is found in several crop species like Brassica, Radish, Nicotiana, and many grass species. It is of two types sporophytic and gametophytic.
- **Male sterility:** In some species, the pollen grains are non-functional. Such condition is known as male sterility. It prevents self-pollination and promotes cross-pollination. It is of three types:, *viz.,* genetic, cytoplasmic and cytoplasmic genetic. It is a useful tool in hybrid seed production.

Study of floral biology and aforesaid mechanisms is essential for determining the mode of pollination of various crop species. Moreover, if selfing has adverse effects on seed setting and general vigour, it indicates that the species is cross-pollinated. If selfing does not have any adverse effect on these characters, it suggests that the species is self-pollinated. The percentage of cross-pollination can be determined by growing a seed mixture of two different varieties together. The two varieties should have marker characters say green and pigmented plants. The seeds are harvested from the recessive (green) variety and grown next year in separate field. The proportion of pigmented plants in green variety will indicate the percentage of out-crossing or cross-pollination.

Significance of Pollination

The mode of pollination plays an important role in plant breeding. It has impact on five important aspects:, *viz.,* (1) gene action, (2) genetic constitution, (3) adaptability, (4) genetic purity (5) transfer of genes.

1. **Gene action:** Gene action refers to mode of expression of genes for various charactors in the population. Gene action is of two types: (i) additive (ii) non-additive (dominance and epistasis). Additive gene action is associated with homozygosity and is more in self-pollinated species, but non-additive gene action is associated with heterozygosity and, therefore, is more in cross-pollinated species.
2. **Genetic constituent:** The breeding populations are of four types: *viz.,* heterozygous, homozygous, heterogeneous and homogeneous. Heterozygous individuals have dissimilar alleles at the corresponding loci and segregate on selfing. Homozygous individuals have similar alleles at the corresponding loci and pure types. They do not segregate

on selfing. The genetically similar population is known as homogeneous. It may be homozygous or heterozygous. Similary genotypically dissimilar populations are known as heterogeneous. Self-pollination leads to homozygosity and cross-pollination results heterozygosity.

3. **Adaptability:** Adaptability refers to stable performance of a variety over a wide range of environmental conditions. The cross-pollinated crop species have better adaptability than self-pollinated species because of more heterozygosity and heterogeneity in the former case. The heterogeneous populations have broad genetic base.
4. **Genetic purity:** Self-pollination maintains the genetic purity of a species and the purity of a variety of self-pollinated crop can be easily maintained for several years. Cross-pollination creater heterozygosity and genetic composition of a cross-pollinated variety change gradually. Hence self-pollination is essential even in cross-pollinated species to maintain the purity in parental lines.
5. **Transfer of genes:** Cross-pollination permits transfer of desirable genes from one species to another. Self-pollination does not permit combining of genes from different source.

Main Features of Cross-pollination

- ☆ The transfer of pollen grain from anther of flowers of a particular plant to the stigma of flowers of another plant of same species (allogamy).
- ☆ Need external agents like birds, insects, winds, water, *etc.*
- ☆ Male and female parts mature at different time (dichogamy).
- ☆ New varieties may be produced by cross pollinating two different varieties of same species. New variations are possible as offspring are healthier and thus adapt themselves to environmental changes.
- ☆ It is also called as allogamy, xenogamy, *etc.*

  e.g. Amaranths, Beat leaf, Spinach, Garden beat, Sugar beet, all Cucurbits, all Cole crops, Radish, Turnip, Carrot Onion, *etc.* Mango (housefly) Citrus, Papaya (wind) Guava (wind), Sopota, Jackfruit, Litchi, *etc.*

Pollinator

Most plants needs in help moving pollen from one flower to the pistil of another flower. Wind moves the pollen for same plants such as grasses like corn. Animal pollinators move pollen for many other flowering plants. Animals that are known to be good pollinators of flowers includes bees, butterflies, hummingbirds, moths, some flies, some wasps and nectar feeding bats, *e.g.,* Anemophillus (by wind), Hydrophillus (by water) and Entomophillus (by insect).

Benefits of Pollinators

Plants benefit from pollinators because the movement of pollen allows them to reproduce by setting seeds. However, pollinators don'ts know or care that the plant benefits. They pollinate to get nectar and/or pollen from flowers to meet their energy requirements and to produce offspring. In the economy of nature the pollinators provide an important service to flowering plants while the plants pay with food for the pollinators and their offspring.

Fertilization

The fusion of male gamete with female gamete is called fertilization. In other words, it is also known as generative fertilization, insemination, pollination, fecundation, syngamy and impregnation which is the fusion of gametes to initiate the development of a new individual organism. The cycle of fertilization and development of new individual is called sexual reproduction. During double fertilization in angiosperm the haploid male gamete combines with two haploid polar nuclei to form a triploid primary endosperm nucleus by the process of vegetative fertilization.

Cross Fertilization

It has been shown in the present volume that the offspring from the union of two distinct individuals especially if their progenitors have been subjected to very different condition have an immense advantage in height, weight, vigour and fertility over the self-fertilized offspring from one of the same parents and this fact is amply sufficient to account for the development of sexual elements that is for the genesis of the two sexes.

Double Fertilization

After pollen is deposited on the stigma it must germinate and grow through the style to reach the ovule. The microspores or the pollen contain two cells the pollen tube cell and the generative cell. The pollen tube cell grows into a pollen tube through which the generative cell travels. The pollen tube is guided by the chemical secreted by the synergids present in the embryo sac and it enters the oval sac through the micropyle of the two sperm cells one sperms fertilizes the egg cell, forming diploid zygote, the other sperm fuses with the two polar nuclei, forming a triploid cell that develops in to the endosperm. Together these two fertilization events in angiosperm are known as double fertilization. After the fertilization is complete, no other sperm can enter.

Difference between Pollination and Fertilization

- Pollination is the transfer of pollen from anther (male organ of the flower) to the stigma (female organ of the flower) of the same or different flower.
- Fertilization occurs once the pollen grain reaches the stigma it produces a pollen tube which grows down through the style reach to the ovary.
- Pollination is prerequisite of reproduction for flowering plants. The word comes from the word pollen which is basically the sperm of the plant. The pollen makes contact with the plants stigma and then the fertilization

process start. The process of pollination was actually discovered in the 18th century by a man named Christian Sprengel.

- ☆ Fertilization is the joining of gametes together in order to produce an offspring. The female egg will be fertilized by a male sperm and this will lead to the creation of a child, be it for animals or for the plants. Basically fertilization occurs not only in animals but in plants as well.
- ☆ Pollination is a process in which flowering plants only undergo and there is transfer of pollen to the plants stigma. The process can be done by the plant itself or through outside agents.
- ☆ Fertilization is basically the joining of sperm and egg. For flowering plants there can be no fertilization if there is no pollination. However fertilization applies to almost every organism in this world.

Chapter 4

Breeding System

Introduction

The possibility of combining sexual and asexual reproduction system in almost all vegetatively propagated horticultural crops is the key strategy to develop new cultivars because hybridization is the best approach to increase genetic diversity and combine traits in an individual. The hermaphrodite flowers may mature at different time facilitating cross-pollination. It is also found as two types.

1. **Protandry:** In it androecium mature first than gynoecium, *e.g.,* Walnut, Coconut, *etc.*
2. **Protogyny:** In it gynoecium mature first than androecium, *e.g.,* Annona, Banana, Plum, Pomegranate, Avocado, Fig, *etc.*

Types of Breeding System

This system is mainly of six types:

A. Apomixis
B. Polyembryony
C. Parthenocarpy
D. Self-incompatibility
E. Sterility
F. Dichogamy

A. Apomixis

The term Apomixis coined by the Winkler (1908) was also study of the asexual reproduction. The term Apomixis are divided in two word where "*Apo*" means

Without and "*Mixis*" means Mingling includes all type of asexual reproduction which stands to replace or to act as substitute for the sexual method. The first discovery of this phenomenon is credited to A.V. Leeuwenhoek as early as in 1719 in citrus seed. Apomixis is widely distributed among higher plant. More than 300 species belonging to 35 families are apomictic. It is most common in gramineae, compositae, rosaceae and rutaceae family crops.

Apomixis refers to the development of embryo without usual process of meiosis and fertilization. The plants developed from apomictic seeds are true to type. In other word apomictic seeds are formed but the embryos develop without fertilization (without sexual process of fusion of male and female gametes) consequently the plant resulting from apomixis is identical in genotype to parent plant. When sexual reproduction does occurs apomixis is termed as facultative and when sexual reproductions are absent, it is called obligate.

"Apomixis refers to the occurrence of an asexual reproductive process in the place of normal sexual process involving reduction division and fertilization. In other words apomixis is a type of reproduction in which sexual organs of related structures take part but seed are formed without union of the gametes, seeds formed in this way are of vegetative in origin." It is mainly of two types.

Apomixis

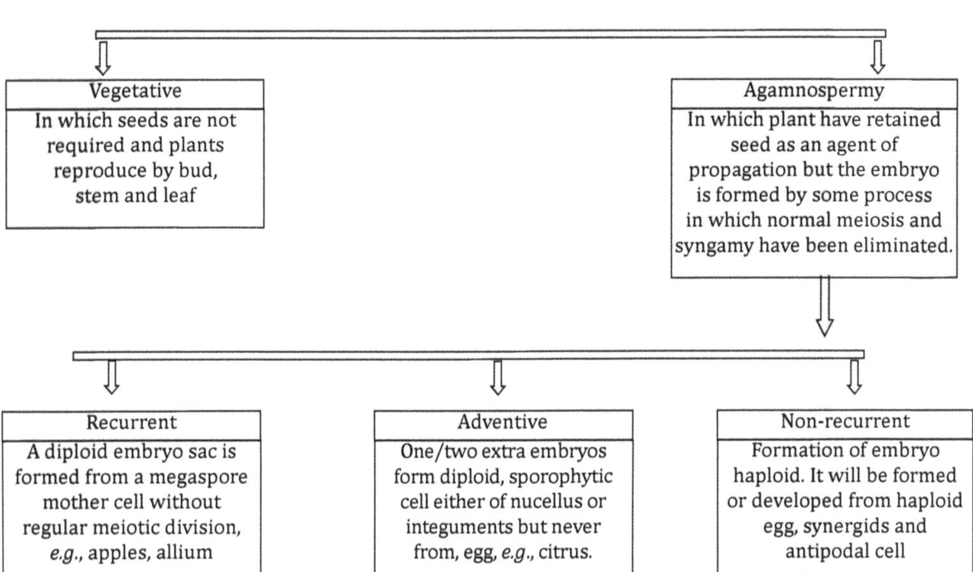

Obligate Apomixis

When apomixis is the only method of reproduction in a plant species, it is known as obligate apomixis.

Facultative Apomixis

When gametic and apomictic reproduction occurs in the same plant, it is known as facultative apomixis.

Classification of Apomixis

1. **Recurrent apomixis:** The embryo sac (female gametophyte) develops from megaspore mother cell whether meiosis is disturbed (sporogenesis failed) or from adjoining cell (megaspore mother cell distintegrates). The egg cell is diploid and embryo develops directly from the diploid egg cell without fertilization. Generally somatic apospory, diploid parthenogenesis and diploid apogamy fall under recurrent apomixis, *e.g., Rubus sps.* (Raspberry), *Malus hupehensis, Malus sikkimensis, Malus sargenti* and *Malus toringoides.*
2. **Non-recurrent apomixis:** The development of embryo takes place from haploid egg cell without fertilization. Such type of apomixis rarely occurs. Generative apospory, haploid parthenogensis, haploid apogamy, androgamy fall under this category.

   Nucellus: In flowering plant the megasporangium is also called the nucellus and the female gametophyte is sometimes called the embryo sac (angiosperm). The megagametophytes produce an egg cell (or several in some group) for the purpose of fertilization.
3. **Adventive embryony:** In this case embryo develops directly from vegetative cells of the ovule, such as nucellus, integuments and chalaza. This is also known as nucellar embryony or polyembryony. In this case more than one embryo develops in a single seed. In the seed both type of embryo develops, *i.e.,* nucellar embryo from nucellar cell and zygotic embryo from egg cell with the result of syngamy. This does not involve production of the embryo sac, *e.g.,* Mango (Olour, Goa, Kurukan, Bappakai, Vellaikolamban, Nileswar Dwarf, Salem, Bellary, Goa Kasargod, Mazagaon, Chandrakaran, *etc.*) and most of *Citrus sps.* except (*Citrus medica* (Citro), *Citrus grandis* (Pummelo) and *Citrus latifolia* (Grapefruit).
4. **Vegetative apomixis:** This is not common in fruit crops. However in some cases like *Poa bulbosa* and *Allium* (Agave) and Grass species vegetative buds or bulbils are produced instead of flower in the inflorescence.

Development of Apomictic Embryo Sac

1. **Apospory:** In Apospory first diploid cell of ovule lying outside the embryo sac and develops into another embryo sac without reduction. The embryo then develops directly from diploid egg cell without fertilization. It involves the development of embryo sac either from the archesporial cell or from the nucellus or from other cell. It is of two types:

- ☆ **Generative or haploid apospory:** If the embryo sac develops from one of the megaspores (n), the process is called haploid apospory, since it cannot regenerate as it is haploid and fertilization fails the process gives rise to non-recurrent apomixis.
- ☆ **Somatic or diploid apospory:** When diploid embryo sac (2n-two megaspores) is formed from nucellus or other cells, the process is termed as somatic or diploid apospory. Since it regenerates, without fertilization it is called recurrent apomixis.

2. **Apogamy:** Development of embryo from synergids or antipodal cell or development of into an embryo within the embryo sac with or without pollination but without fertilization is termed as apogamy. In this case, there could be found haploid and diploid apogamy depending upon the ploidy level. Diploid apogamy is recurrent type and haploid apogamy is non-recurrent types, *e.g.,* Diploid- *Allium, Iris.*
3. **Parthenogenesis:** It can be defined as development of embryo from egg cell with or without pollination but without fertilization. Depending upon ploidy level of egg cell parthenogenesis can be haploid (when the embryo develops from a diploid or sterile egg cell are non-recurrent) or diploid (sometimes embryo sac develops without reduction division, such embryo sac and the entire cell within it, are diploid are recurrent), *e.g.,* Banana, Pineapple, Mango, *etc.*
4. **Androgamy:** Development of embryo from male gametes inside or outside the embryo sac is known as androgamy. Since the cell are haploid in nature they come under non-recurrent types.
5. **Megaspore mother cell:** Megasporocyte is a diploid cell in plants in which meiosis will occur resulting in the production of four haploid megaspore. At least one of the spores develops in to haploid female gametophytes. The megaspore mother cell arises within the mega sporangium tissue.
6. **Endosperm:** It is the tissue produced inside the seed of most of flowering plants following fertilization. It surrounds the embryo and provides nutrition in the form of starch, though it can also contain oil and protein. This can be source of nutrient for human diet.
7. **Antipodal cell:** These are nutritive in function and they nourish the embryo sac.
8. **Synergids:** These cells are two specialized cells that lie adjacent to the egg cell in the female gametophyte of angiosperm and play an essential role in pollen tube guidance and function.

Genetics of Apomixis

The apomictic condition is recessive to sexuality although polyploid apomictics show tendency towords dominance. However this recessiveness is not usually due to a monogenic difference. A successful apomictic cycle is the result of an interaction of many genes which tend to break the hybridization. It is only in the relatively simple type of apomixis like adventive embryony and vegetative reproduction that simple genetic behaviour can be expected.

Delection of Apomixis

From an intensive screening of a large number of plant varieties/hybrids which involves a careful and systematic tracing of steps for the development of embryo sac and embryo through microtomy of ovule, right from megaspore to embryonic development. It should, however, be noted that it is only the recurrent apomixis (diploid) form apospory/parthenogenesis, apogamy and adventive embryony and the vegetative propagation which are beneficial for plant breeding purposes. Non-recurrent apomixis is of academic importance only.

Maintenance and Transfer of Apomixis

Once an apomictic plant is detected its inheritance pattern may be studied through crossing a few sample flowers with the pollen obtained from normal plants and observing segregation pattern in F_2 and subsequent generation. The remaining flowers may thoroughly be checked and seed collected on maturity. The true apomictic plant well automatically produces mother apomictic progenies which can be maintained without difficulty.

Advantage of Apomixis in Plant Breeding

Apomictic tends to conserve the genetic structure of their carrier and are also capable of the maintaining the advantages of heterozygote generation after generation. Therefore such a mechanism might offer a great advantage in plant breeding where genetic uniformity maintained over generations for homozygosity (in varieties of selfers) and heterozygosity (in hybrid of both selfers and out breeders) is the choicest goal. Apomixis may also effect an efficient exploitation of maternal influence, if any reflecting in the resultant progenies early or delayed because perpetuation of the only maternal properties due to prohibition of the fertilization. Maternal effect is most common in horticultural crops.

- ✰ Faster multiplication of genetically uniform individual can be done without any problem of segregation.
- ✰ Heterosis/hybrid vigour can permanently be fixed in plant.
- ✰ Efficient exploitation of maternal effect (if present) is possible for generation to generation.

Disadvantage of Apomixis in Plant Breeding

- ✰ Apomixis is nuisance when the breeder desires to obtain sexual progeny, *i.e.,* self or hybrid.

Amphimixis: Amphimixis is a kind of sexual (true) reproduction in which both processes meiosis and fertilization occur. In meiosis chromosome number is reduced to half and in fertilization chromosome number restores (2n).

Apospory: Formation of gametophyte directly from sporophyte (nucellus or integuments) without meiosis is called apospory. Such aposporous gametophyte will be diploid, *e.g.,* Malus, Crepsis, Ranunculus, *etc.*

Apogamy: The formation of sporophyte directly from the gametophyte except egg without fertilization is called as apogamy and sporophyte is haploid, *e.g.,* Orchid, Masculata, *etc.*

Adventive embryony: In this case embryos develop directly from vegetative cells of the ovule such as nucellus, integuments and chalaza and do not involve production of embryo sac, *e.g.,* Mango, Citrus, *etc.*

B. Polyembryony

The term Polyembryony was first coined by Antonie van Leeuwenhoek (1719) especially in the orange commonly found in citrus. The term Polyembryony is divided in two word where "*Poly*" means Many or More than one and "*Embrony*" means Embryo.

"The occurrence of more than one embryo in the ovule is called as polyembryony which is very common in gymnosperm than the angiosperm because in gymnosperm there are many archegonia which after fertilization may produce more than one embryo but in angiosperm rarely found". It is mainly of two types:

1. **True polyembryony:** Extra embryos develop within the embryo sac in ovule within same embryo sac and same ovule.
2. **False polyembryony:** Extra embryo develops outside of the embryo sac in nucellus or integument.

It is the occurrence of more than one embryo in a seed which results in the emergence of multiple seedlings. The additional embryos results from the differentiation and development of various maternal and zygotic tissues associated with the ovule of seed. It is considered as desirable character in citrus, mango, jamun, rose, apple, almond, *etc.* to obtaining true to type planting materials. In the usual process of the plant reproduction egg cell (female) of ovule fertilizes with the male gamets which in turn culminates in embryo (zygotic) and on the other hand in some of crops somatic cell of ovule (usually nucellus) start producing embryo without fertilization and this results in the occurrence of more than one embryo in the seed (polyembryony).

Where on the formation of embryo without sexual process is called apomixis and seed is called apomictic seed whereas, condition is called polyembryony. In the most of cases sexual embryo and nucellar embryo develop simultaneously which is known as facultative polyembryony and polyembryony in horticultural crops is associated with nucellar embryony because it frequently results in polyembryony from seed in which multiple seedlings germinate. About 244 species of 145 genera belonging to 59 angiosperm families are reported to exhibit polyembryony. It is mainly found with more than 40 or 40 embryos in citrus species mainly Satsuma mandarin or tangerine (*Citrus unshiu*). It is the seedless fruit and loose skin or loose jacket found.

The phenomenon in which more embryos are present within a single seed is called polyembryony. It may be due to nucellar embryony (citrus) and development of more than one nucleus within the embryo sac (in addition to egg embryo during the early stage of development) leading to multiple embryos (conifers). In citrus, it varies from species to species. In rough lemon it varies from 3-5 embryos. In mango, it varies from 2-10 embryos with germination percentage 40-85 per cent. Polyembryonic seedling can be identified from its true seedling by their uniformity and vigorous growth while the seedling arising from fertilizated embryo will be weak. In mango polyembryony is determined by single dominant gene.

Classification of Polyembryony

(A) On the Basis of Characters

- ☆ **Simple:** Fertilization of more than one egg.
- ☆ **Adventive:** It is process of formation of extra embryo by sporophytic buds.

(B) On the Basis of Frequency

- ☆ **Sterility:** Plant species in which the frequency of multiple embryos is less than 7 per cent.
- ☆ **Nearly polyembryonic:** In case of nearly monoembryony plant species, frequency of polyembryony varies between 6-10 per cent.
- ☆ **Polyembryonic:** The percentage of multiple seed is more than 10 per cent.

(C) On the Basis of Genetic Composition

- ☆ **Gametophytic:** Multiple embryos arise from the gametic cells of the embryo sac (synergids, antipodal) after or without fertilization. In this case haploid embryos are formed.
- ☆ **Sporophytic:** When multiple embryos arise either from zygote or from the sporophytic cells of ovule (nucellus, integuments) and the resulting embryos will be diploid, *e.g.,* Citrus, Mango, Jamun, Rose, Apple, Almond, Peach and Onion.

Citrus is most polyembryonic crop. It is most important group exhibiting this traits except *Citrus grandis* (Pummelo), *Citrus latifolia* (Tahiti lime) and *Citrus medica* (Citron). All other species show monoembryony.

But in mango, major polyembryonic varieties are Bappakkai, Chandrakaran, Kensington, Kitchner, Kurukkan, Muvandan, Mylepelion, Nekkare, Olour, Peach, Prior and Starch. Mango varieties of economical importance are monoembryonic. In polyembryonic cultivars the vigorous growth of nucellar embryos inhibits the growth of zygotic embryo and causes its degeneration prior to seed maturation.

(D) On the Basis of Development

- ✫ **Zygotic or suspensor polyembryony:** Cleavage of the apical cells of the globular or filamentous polyembryony produced by the zygote may result in two or more embryos in a seed. The multiple embryos that arises from polyembryonal or suspensor cells are diploid, *e.g.,* Orchids.
- ✫ **Nucellar or adventive polyembryony:** Diploid nucellus or integument cells form embryos, *e.g.,* Citrus, Opuntia and Magnifera.

(E) On the Basis of Introduction

- ✫ **Induced:** It includes cases of experimentally induced polyembryony.
- ✫ **Spontaneous**: It includes all cases of naturally occurring.

(F) On the Basis of Process

- ✫ **Cleavage:** In this type a single fertilized egg gives rise to number of the embryos.
- ✫ **Simple:** In this type number of embryos develop as a results of the fertilization of several archegonia.
- ✫ **Rosette:** In some gymnosperms (*e.g.,* a few species of *Pinus*) additional embryos develop from rosette cells and this type of polyembryony has been term as rosette polyembryony.

Archegonia

It is multicellular structure or organ of the gametophyte phase of certain plant producing and containing the ovule or female gamete. The corresponding male organ is called antheridium. The archegonia have a long neck canal or venter and swollen base. Archegonia are typically located on the surface of plant thallus although in the hornworts they are embedded.

Degree of Polyembryony

Number of embryos varies with species and varieties. Seed of highly polyembryony species contains even more than 8 embryos. However the occurrence of 2-4 embryos is common. The intensity of occurrence of seeds containing multiple embryos may range up to 70 per cent in highly polyembryony species. The emergence of seedlings is not directly correlated with the number of embryos usually less number of seedlings emerged with respect to the number of embryos. The intensity of occurrence of multiple seedling increases with the increase in the number of embryo. In citrus the occurrence of one seedling/seed was more than 50 per cent whereas, the intensity of emergence of two seedlings was 36 per cent.

Identification of Nucellar Seedlings

In case of facultative polyembryony both sexual and nucellar embryos are formed and they germinate in the form of seedling between them only nucellar seedling processes the genetic make up to mother plant and are true type, but its morphological identification is difficult. However nucellar embryos can be distinguished from zygotic embryo by their lateral position in the embryo sac, irregular shape and lack of suspensor. The zygotic seedlings are frequently smaller than nucellar seedlings but variation enough in height or size alone could not be used as criteria for selection.

- In mango zygotic seedlings are vigorous and possess big cotyledons. Visual identification of nucellar seedling is not an effective method thus it becomes often necessary to grow seedling for 4-5 years to fruiting before nucellar seedling can be distinguished from zygotic ones.
- In citrus zygotic seedlings can also be identification if one variety is crossed with trifoliate orange, as the zygotic seedlings will have trifoliate leaves for a more rapid advance in the propagation programme. It is necessary to identify zygotic seedling at an early stage (as a seedlings) and for this purpose biochemical and molecular methods are used.

Significance of Polyembryony

1. The adventive embryos develop from nucellar embryo provide uniform seedlings for root stock which yield consistent results in fruit production.
2. The nucellar seedling as a root stock has a tap root and therefore, develops a better root system.
3. The nucellar seedlings show a restoration of the vigour lost after repeated vegetative propagation.
4. Polyembryony has ecological significance as it increases the probability of survival under varied conditions.
5. Nucellar polyembryony is the only practical approach to raise virus free clones of polyembryonic *Citrus* sps. in nature.
6. Disease free plant can also be obtained through nucellar embryo culture.

Importance of Polyembryony

- Multiple embryos within a seed can be haploid, diploid or triploid.
- Adventive embryos are useful in agriculture and horticulture because they are genetically uniform and generally disease free.
- Environmental factors are often responsible for induction of polyembryony.
- Application of PGRs has not proved successful in inducing polyembryony.

Reason of Polyembryony

- ☆ Cleavage of polyembryony.
- ☆ Development of many embryos from other cells of embryo sac except egg (nucellus, integument).
- ☆ Formation of many embryos from the same embryo sac and ovule.
- ☆ Formation of many embryos from sexual structures outside of embryo sac.

C. Parthenocarpy

The term Parthenocarpy is derived from in two Greek word where the '*Parthenos*' means Virgin (not fertilized) and '*Carpic*' means Pertaining to fruit (making fruit). The term parthenocarpy was first coined by the Null (1902). The phenomenon of the formation of fruit without fertilization is known as parthenocarpy such fruits are seedless. Natural parthenocarpy is found in grape, pineapple and banana and parthenocarpy can be induced by the treatment of hormones such as auxin and GA_3. Generally parthenocarpy arises due to mutation and hybridization, varieties of grapes and cucumber have resulted from bud mutation. Unpollinated flowers are exposed with phytohormones (IAA, Auxin and GA_3) on stigma to produce seedless fruit. In case of fruit formation without pollination and fertilization, the fruit resembles a normally produced fruit but is seedless. For example, cucumber, pineapple, banana, grape, orange, grapefruit, persimmon and breadfruit.

The naturally occurring parthenocarpy (seedless parthenocarpic fruit) can be induced in non-parthenocarpic varieties and in naturally parthenocarpic varieties out of season by a type of artificial pollination with dead or altered pollen or by pollen from a different type of plant. The application of synthetic growth substances in paste form by injection or spraying of auxin and GA_3 can also give rise to parthenocarpy.

FLOW CHART OF PARTHENOCARPY

No pollination and fertilization

Fruit formation without seed

Natural condition developing embryo

It self produces hormone

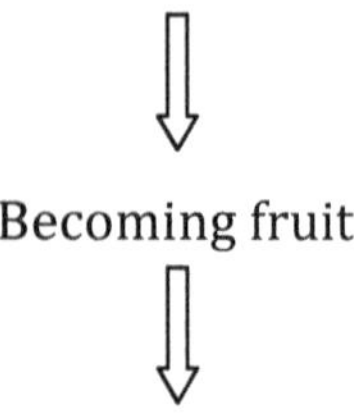

Unpollinated flower with phytohormones

In botany and horticulture parthenocarpy (literally meaning virgin fruit) is the natural or artificial induced production of fruit without fertilization of ovules which makes the fruit seedless. Stenospermocarpy may also produce apparently seedless fruit but the seed are actually aborted while they are still small parthenocarpy or stenospermocarpy occasionally occurs as a mutation in nature if it effects every flower the plant can no longer sexually reproduce but might be able to propagate by apomixis or by together vegetative means. Parthenocarpic fruit are usually seedless but need not be always.

Types of Parthenocarpy

1. **Vegetative parthenocarpy:** If a fruit develops even without the stimulus of pollination then the phenomenon is referred to as vegetative parthenocarpy (apomictic), *e.g.*, Banana, Japanese persimmon, Apple, Pineapple and Cucumber.
2. **Stimulative parthenocarpy:** If a fruit develops from the more stimulus of the pollination (but without fertilization) the phenomenon is known as stimulative parthenocarpy. The female flowers of triploid watermelon require the pollen grains of diploid varieties to develop into a seedless fruit. Diploid pollen grains give a stimulus to the ovary of the plant when self pollinated which results in the development of parthenocarpic fruit due to the stimulation provided by pollen hormones, *e.g.*, Thompson Seedless variety of Grapes and Papaya, Watermelon, *etc.*
3. **Stenospermocarpy:** In Black Corianth or Thompson Seedless variety of grape pollination and fertilization take place but the embryo gets aborted subsequently resulting in seedlessness. This phenomenon of development of seedless fruit is referred to as stenospermocarpy, *e.g.*, Watermelon, Grape *etc.*

   The term stenospermocarpy was coined by the Stout (1936). The problem of unfruitfulness due to pollination failure, sterility and incompatibility may not arise if a fruit develops parthenocarpically and the grower is assured of good crop (*e.g.*, Banana). One drawback with the seedless fruit is that they are usually small in size (Black Corianth variety of Grape).
4. **Artificial parthenocarpy:** Some parthenocarpy varieties have been developed as genetically modified organisms when sprayed on flowers of the PGR (GA, auxin) could stimulate the development of parthenocarpic

fruit that is termed as artificial parthenocarpy, *e.g.*, Pineapple, Banana, (IAA, GA_3), Grape, Citrus (IAA), Tomato (IAA GA_3) and Brinjal (2,4-D, IAA).

Induction of Seedlessness in Fruits

The seedlessness can be induced by the following methods.

- ✰ **Use of growth regulators:** Application of GA_3 at 8000 ppm in lanolin paste on the cut end of the style of the emasculated flower of guava resulted in the development of seedless fruits, similarly seedlessness in loquat was induced by spraying GA_3 100-200 ppm on the emasculated flowers.
- ✰ **Changing the ploidy level:** It was first demonstrated in Japan by developing a triploid watermelon (2n=33) by crossing tetraploids x diploid varieties.

Parthenogenesis

The formation of embryo directly from the egg without fertilization is called as parthenogenesis. In some plants fruit develops parthenocarpically, still they produce viable seeds (*e.g.* Mangosteen and Strawberry). This phenomenon is referred to as parthenogenesis. The seedlings of such fruit are genetically uniform. In certain cases seed develops parthenogenetically but they are non-viable (*e.g.* Apple). When female flower of jackfruit are pollinated with the pollen grains of bread fruit, seeds are formed in jackfruit but they did not germinate as they are non-viable. It has two types:

(A) Haploid Parthenogenesis

1. When embryo develops from haploid egg as Datura, *Solanum nigrum*, Oenothera, *etc.*
2. Embryo is haploid and sterile.

(B) Diploid Parthenogenesis

1. When embryo develops from diploid egg and formed embryo is diploid as aposporous plant, *e.g.*, Hieracium, Ranunculus, *etc.*
2. Sometimes embryo sac develops without reduction division such embryo sacs and all the cells within it are diploid, *e.g.*, Grass like, Taraxaccum, *etc.*

Types of Parthenocarpy

1. **Natural or obligatory:** Development of seedless fruit due to genetic sterility (requires vegetative method of propagation), *e.g.*, Banana, Japanese persimmon, Pineapple, *etc.*
2. **Artificial or facultative:** Production of seedless or seeded fruit due to environmental stimulation. Parthenocarpy in tomato is due to single recessive gene (pat-2), *e.g.*, Grape, Tomato, Mutants, Citrus cultivars, Cucumber and Watermelon.

Seedless Cultivars in Citrus

- ☆ Mandarin – Clemenules
- ☆ Sweet orange – Washington Naval Orange
- ☆ Tahiti lime – Due to triploid
- ☆ Oroblanco – *Citrus grandis* x *Citrus paradisi*
- ☆ Satsuma mandarin – *Citrus unshiu*

D. Self-Incompatibility (SI)

The term Self-incompatibility was originally coined by Stout in 1917. Koelreuter in the middle of 18th century first reported self-incompatibility in *Verbascum phoeniceum* plants. Later on numerous cases of self-incompatibility in flowering plants were reported which were reviewed by East (1940). It is the failure of pollen tube to penetrate the full length of style and effect fertilization.

It refers to failure of pollen to fertilize the same flower or other flower of the same plant or the failure of pollen to fertilize. The barrier between pollination and fertilization in angiosperms is because of the self-incompatibility a genetically controlled phenomenon. If a plant does not set seed when pollinated with its own functional pollen, but exhibits normal seed setting when cross pollinated, it indicates presence of self-incompatibility, thus self-incompatibility refers to the inability of functional male and female gametes of the hermaphrodite flowers to set seed on self-pollination.

Main Features of Self-Incompatibility

1. It is an important outbreeding mechanism which prevents autogamy (self) and promotes allogamy (cross).
2. Self-incompatibility species do not produce seed on self-pollination but lead to normal seed set on cross-pollination.
3. It maintains high degree of heterozygosity in a species due to outbreeding and reduces homozygosity due to elimination of inbreeding or selfing.
4. It results due to morphological, genetic, physiological and biochemical causes. It is not under simple genetic control.
5. Self-incompatibility reaction can operate at any stage between pollination and fertilization.
6. It has been reported in about 300 species belonging to 70 families of angiosperm including in several crop species.

Self-incompatibility, non fruitfulness or failure of fruit setting may occur due to:

1. Pollen grains fail to germinate on the stigma.
2. Pollen grains germinate but pollen tube fails to enter inside the stigma and style.

3. Sometime pollen tube enters inside the style but growth is very slow.
4. Pollen tube enters the ovule but there is no fertilization due to degeneration of egg cells.
5. Sometimes fertilization is effected but embryo degenerates at very early stage.

Classification of Self-Incompatibility

Self-incompatibility has been classified by Lewis 1954 on the basis of -

1. ***Flower Morphology***
   (a) **Heteromorphic- Distyly** (Style and stamens are short and long)
   Tristyle (Style and stamens are short, medium and long)
   (b) **Homomorphic- Sporophytic** (Genotype of pollen producing plant)
   Gametophytic (Genetic constitution of gametes)
2. ***Genes involved= Monoallelic (Controlled by single gene)***
   Diallelic (Governed by two genes)
   Polyallelic (Governed by several genes)
3. ***Site of expression= Stigmatic (SI gene expresses on stigma)***
   Stylar (SI gene expresses on style)
   Ovarian (SI gene expresses on ovary)
4. ***Pollen cytology= Binucleate (Pollen grain have two nuclei)***
   Trinucleate (Pollen grain have three nuclei)

(A) Heteromorphic System

When self-incompatibility is associated with differences in floral morphology it is known as heteromorphic system. In this system self-incompatibility results due to the differences in the length of style and stamen.

1. Distyle

It refers to two types of styles (short and long) and stamen (low and high). This system operates in the family Primulaceae. In Primula there are two types of flower.

(a) **Thrum:** This type which has short style and high anthers and long stamen.
(b) **Pin:** This type with long style and low anthers and short stamen.

The crosses are compatible between pin x thrum or thrum x pin but not between pin x pin and thrum x thrum flower. Later on it was discovered that incompatibility barrier between pin x pin and thrum x thrum is governed by a single gene S which behaves in a Mendelian fashion. It is considered that thrum is heterozygous (Ss) and pin is the homozygous recessive (ss). Thus thrum is dominant over pin crosses between thrum and pin produce thrum and pin in 1:1 ratio in F_1

as given below (Allele S is dominant over s). The incompatibility reaction of pollen is determined by the genotype of the plant producing them.

Pin x Pin	ss x ss	(Incompatible)
Thrum x Thrum	Ss x Ss	(Incompatible)
Thrum x Pin	Ss x ss	(1Thrum: 1 Pin)
Pin x Thrum	ss x Ss	(1Thrum: 1 Pin)

For example, sweet potato is genotypically S and s but all of them would be S phenotypically.

2. *Tristyle*

When style has three different positions it is referred to as tristyle. In tristyle anthers and style have three positions in flowers, *viz.,* short, medium and long. Tristyly is common in the family Lythraceae (*Lythrum salicaria*). Three positions of style are genetically controlled by two genes (Ss and Mm). Short style may have Ssmm, SsMm or SsMM genotype, medium style ssMm or ssMM genotype and long style has ssmm genotype.

(B) Homomorphic System

In homomorphic system self-incompatibility results due to physiological causes rather than differences in flower morphology. In this system the plants do not have differences in the length of style and stamen or other floral parts. The incompatibility reaction of pollen may be controlled by the genotype of plant on which it is produced (sporophytic control) or by its own genotype (gametophytic control).

- ☆ The pollen grains do not germinate on the stigma of same flower. If they germinate the pollen tube fails to penetrate the stigma, *e.g.,* Cabbage and Radish.
- ☆ The pollen grain may germinate but there is retardation of pollen tube growth.
- ☆ In some cases there is slow rate of pollen tube growth and it rarely reaches the ovary in times to effect self-fertilization.
- ☆ In some cases pollen tube growth may be normal but is does not release the male gamete.

1. *Gametophytic System*

When the self-incompatibility is controlled by the genetic constitution of gametes it is known as gametophytic self-incompatibility system. This system was first discovered by East and Mangels dorf (1925) in *Nicotiana sanderae* (Tobacco), *e.g.,* Potato, Tomato, and Loquat, Petunia (1-2 Locus) Sugerbeet (4 or more locus), *etc.*

Main features of gametophytic system

1. Self-incompatibility is majority of species is governed by a single gene S, which has large number of multiple alleles.

2. In this system allele have individual action in the style without interaction.
3. Pollen grains are unable to germinate or function on a pistil having similar allele as that of pollen. The pollen tube growth is usually inhibited in the style or ovary.
4. This system gives rise to three types of pollination:
   - ☆ **Fully incompatible (S_1S_2 x S_1S_2)-** In which both alleles are common in pollen and ovule.
   - ☆ **Partial self-incompatibility/Half the pollen is compatible (S_1S_2 x S_1S_3)-** In which one allele is different.
   - ☆ **Fully compatible/fertile (S_1S_2 x S_3S_4)-** When both alleles differ in pollen and ovule.
5. This system permits recovery of male parent only in the partially fertile crosses which are obtained when one alleles differs in the cross *viz.*, S_1S_2 x S_1S_3. This cross would give rise to S_1S_3 and S_2S_3 progeny.
6. The bio-chemical substance which is associated with the incompatibility response of the pollen develops very late during pollen formation in gametophytic system (Binucleate pollen in plant species).

The incompatibility reaction of pollen is determined by its own genotype and not by the genotype of the plant on which pollen is produce. S_1S_2 the pollen carrying S_1 or S_2 allele would behave as S_1 if S_1 is dominant and if there is no dominance both will behave as S_1 plus S_2.

S_2 pollens produced from S_1S_2 parent

⇓ (Never germinates)

S_1S_3 (stigma)

S_1S_2 pollen

⇓ (Never germinates)

S_1S_4, S_1S_5, S_2S_3, S_2S_9
(stigma)

S_1S_2 pollen

⇓ (100 per cent germinate)

S_3S_4, S_3S_9

(1) Sporophytic System

When the self-incompatibility is governed by the genotype of pollen producing plant (sporophyte), it is called sporophytic system. This system was first discovered by Hughes and Babcock in *Crepis foetida* and Gerstel (1950) in *Parthenium argentatum* (Guayule), *e.g.*, Radish, Cabbage, Cauliflower, Cosmos, Mango, *etc.*

Main features of Sporophytic System

1. Self-incompatibility is governed by a single gene S with multiple allele more than 30 alleles are known in *Brassica oleracea.*
2. In general the number of S alleles is considerably larger in gametophytic system than in sporophytic system.
3. The incompatibility reaction of pollen is governed by genotype of the plant on which the pollen is produced and not by the genotype of the pollen.
4. The alleles may show dominance, co-dominance or competition, individual action or interaction in either pollen or style as per the allelic combination involved.
5. This system exhibits inhibition of pollen germination or pollen tube growth on the stigma of same flower.
6. The sporophytic system contains a form of dominance in which S_1 is dominant over all other alleles S_2 is dominant over all except S_1 and S_0 on (S_1>S_2>S_3>S_4). In this system crosses between different genotypes are either fully fertile or completely sterile crosses of S_1S_2 (female) x S_1S_3 (male) parent.
7. Pollen grains form both heterozygous or homozygous, plants react in a similar fashion due to dominance effect of male parent (some species trinuclate pollen), *e.g.*, S_1S_2 x S_1S_2 (S_1 phenotype) S_2S_4 x S_2S_4 (S_2 phenotype).

Significance of Self-Incompatibility

Self-incompatibility effectively prevents self-pollination as a results it has a profound effect on breeding approach and objective.

- ☆ In self-incompatibile fruit trees it is necessary to plant two cross compatible varieties to ensure fruitfulness. The further cross-pollination may be poor in adverse weather condition reducing fruit set. Therefore, it would be desirable to develop self-fertile types in such case.
- ☆ Some breeding schemes are also suitable for the development of hybrid variety, *etc.*
- ☆ It is useful in hybrid seed production.
- ☆ Discourages the self-pollination and promotes the cross-pollination (out breeding).

Comparison between GSI and SSI

Gametophytic System	*Sporophytic System*
1. Self-incompatibility is controlled by genetic constitution of pollen.	1. Self-incompatibility is controlled by genotype of pollen producing plant (sporophytic).
2. Incompatibility is governed by a single S gene with multiple alleles.	2. Self-incompatibility is also governed by a single S gene with multiple alleles.
3. Alleles have individual action in the style without interaction.	3. Alleles may show dominance, individual action or interaction in either pollen/style.
4. The pollen tube growth is usually inhibited in the style or ovary.	4. Pollen germination or pollen tube growth in inhibited on stigma.
5. Plant species belonging to this system have binucleate pollen grain.	5. Plant species belonging to this system have trinucleate pollen grains.
6. Reciprocal differences are not observed.	6. Reciprocal differences are observed.
7. Recovery of only male parent is possible from cross.	7. Recovery of both male and female parents possible from cross.
8. Crosses may be sterile partially fertile/fully fertile.	8. Crosses would be either fully sterile or fully fertile.
9. Such self-incompatibility can be overcome by polyploidy.	9. This cannot be overcome polyploidy.
10. *e.g.*, Potato and Tomato.	10. *e.g.*, Radish, Cabbage, Cauliflower, *etc.*

Mechanism of Self-Incompatibility

There are two different types of events which constitute the basis of self-incompatibility system.

1. The stimulation of unlike genotypes.
2. The inhibitioned of like genotypes.

Thus two hypothesis have been proposed to explain the mechanism of self-incompatibility in plants.

1. Complementary hypothesis.
2. Oppositional hypothesis.

1. Complementary Hypothesis

This hypothesis was proposed by Batmen in 1952. In this self-incompatibility results due to absence of stimulation by the pistil on pollen growth in the like genotype ($S_1S_2 \times S_1S_2$). In other word self-incompatibility results due to absence of substances in the pistil or pollen which is essential for pollen tube penetration on selfing the pollen and or the pistil fail to produce the substance which is essential for the pollen to germinate or for pollen tube growth in the style and the ovary.

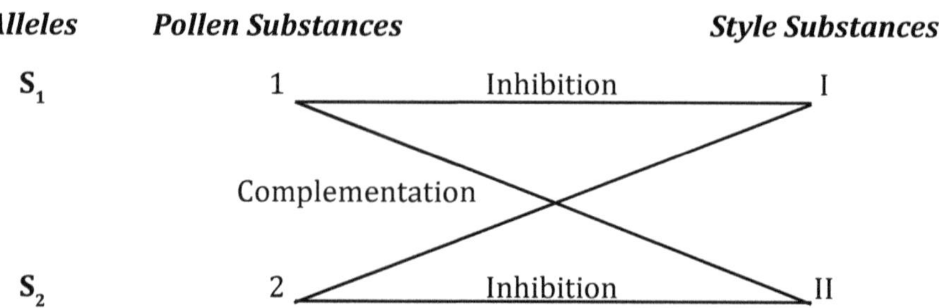

2. Oppositional Hypothesis

This hypothesis states the interaction between like alleles (S_1S_2 x S_1S_2) leads to production of inhibitor which inhibits the growth of pollen tube in the pistil. In other words as a result of interaction between like alleles a substance is produced in pollen and pistil which has the property to interfere with the normal metabolism of the pollen grain or the pollen tube. The inhibitor can act in three ways-

1. It may inhibit an enzyme or auxin necessary for pollen tube growth.
2. May block pollen tube membrane.
3. May inhibit an enzyme necessary for the penetration of style.

Site gene expression: Based on the various phenomenon observed during pollination and fertilization it can be expressed only at three different locations in the flowers.

- **Pollen-stigma interaction:** This interaction occurs just after the pollen grains reach the stigma and generally it prevents pollen germination. In gametophytic system, stigma surface is plumose having elongated receptive cells and is commonly known as wet stigma. Incompatibility reaction occurs at a later stage. There is clear cut serological difference among the pollen grains with different S genotypes and such differences have not been observed in sporophytic system. In sporophytic system stigma is papillate and dry and covered with a hydrated layer of proteins known as pellicle. There are striking differences in the stigma antigens related to the S allele composition within few minutes of reaching the stigmatic surface the pollen releases exine exudates which are either protein or glycoprotein in nature. This exudate induces immediate cellose formation in papillae (which are in direct contact with the pollen) of incompatible stigma. Thus in the sporophytic system stigma is the site of incompatibility reaction.
- **Pollen tube-style interaction:** In most of the gametophytic system pollen grains germinate and pollen tube penetrate the stigmatic surface. But in the incompatibility combination the growth of pollen tube is retarded within the stigma.

☆ **Pollen tube-ovule interaction or ovarian inhibition:** In some cases pollen tube reaches the ovule and affects fertilization. However, in incompatible combination embryo degenerates at early stage of development.

Methods of overcoming self-incompatibility: Following methods can be used for bringing partial fertility by temporarily suppressing the incompatibility reaction.

Bud pollination:-Pollination of immature bud or immature non-receptive stigma with application of mature pollens has been successfully used in production of large quantity of self seed both in gametophyte and sporophytic system in *Brassica, Nicotiana* and *Petunia*. This method is used to develop inbred lines for hybrid seed production in *Brassica* spp. The bud pollination should be done 2-4 days before flower opening or 1-2 days prior to anthesis.

Delayed pollination: It leads to self-fertilization in some self-incompatibility species where pollination of aged pistil is done several days after maturity with normal incompatible pollen and this result in some degree of self-fertilization in *Brassica* and *Linum*.

1. **Late season pollination:** Self-pollination at the end of flowering season also leads to seed set in some species like *Nicotiana* and *Petunia*.
2. **Irradiation:** Irradiation of style with X-rays or Gamma rays for single locus gametophytic incompatibility immediately before selfing resulted in breakdown of self-incompatibility in *Petunia*. Where Gamma rays increase seed setting upon selfing in tomato and *Nicotiana*. However these effects were observed in one generation and were lost in next generation. If S_1 is mutated to S_2 it will be compatible with both S_1 and S_2.
3. **High temperature:** Treatment of style (pistils) at temperature ranging from 30°C to 60°C leads to breakdown of self-incompatibility in many plant species like *Malus, Pyrus, Prunus, Trifolium, Lycopersicon, etc.*
4. **Surgical technique/mutilation of style:** Removal of stigma in species having stigmatic self-incompatibility results in certain amount of seed production in some crops. Mutilating by steel wire brush during pollination is also useful to break down self-incompatibility in *Brassica oleraceae*.
5. ***In-vitro* fertilization:** Placing of pollen grain in direct contact with ovules resulted in breakdown of self-incompatibility in many crops. Self seed were obtained in *Petunia axillaris* by this method. The restriction of interference from the stigma or the style or the ovary leads to breakdown of self-incompatibility.
6. **Double pollination:** Incompatible pollen applied as mixture with compatible pollen gives desirable results sometimes. Arora and Singh (1988) observed that in low chilling Plum and Peach cultivars method of killing the mentor pollen was not helpful in overcoming incompatibility

barriers, however frozen and thawed mentor pollen (one which, if alive would be fully compatible with style receiving it) improved fruit set in both intra and inter specific incompatibility.

Advantage of Self-incompatibility

- Where male sterility is nonexistent, self-incompatibility can alternatively facilitate the production of F_1 hybrids.
- Self-fertility can be induced temporarily or permanently by mutation of S alleles to S_1 through artificial irradiation in clonally propagated orchard species like cherry and apple and temperate fruits.
- Seedless varieties such as pineapple, grape, *etc.* can be evolved if self-incompatibility is present.

Disadvantage of Self-Incompatibility

- Variations in seed set due to poor fertility.
- Poor preservation of genetic purity of improved varieties since cross-pollination is non-restricted.
- Difficulties in development and maintenance of homozygous lines (inbred) which can be utilized for hybridization.
- Uneven quality of fruit due to mixed planting of different varieties based on their cross-compatibility.

Limitation of Self-Incompatibility

- It is very difficult to produce homozygous inbred lines in a self-incompatibility species. Bud pollination has to be done to maintain the parental lines.
- It is affected by environment factor such as temperature and humidity. Incompatibility is reduced or brokendown at high temperature and humidity for maintaining parental lines.
- Sometimes bees visit only one parental line in the seed production plot resulting in sib mating, *e.g., Brussels sprouts.*

Pollination pattern and incompatibility: Self-incompatible fruit cultivars or species need cross-pollination for seed or fruit set which includes pollen hydration and germination, pollen tube growth into the style to the ovary, entry into the ovule and embryo sac and release of its sperms. The germination of pollen grain and its penetration into the stylar tissue to reach embryo sac depends upon acceptability by the pistil which is selective in nature, *e.g.,* Aonla, Apple, Ber, Citrus, Grape, Guava, Jackfruit and Mango.

POLLEN - STIGMA - OVULE INTERACTION

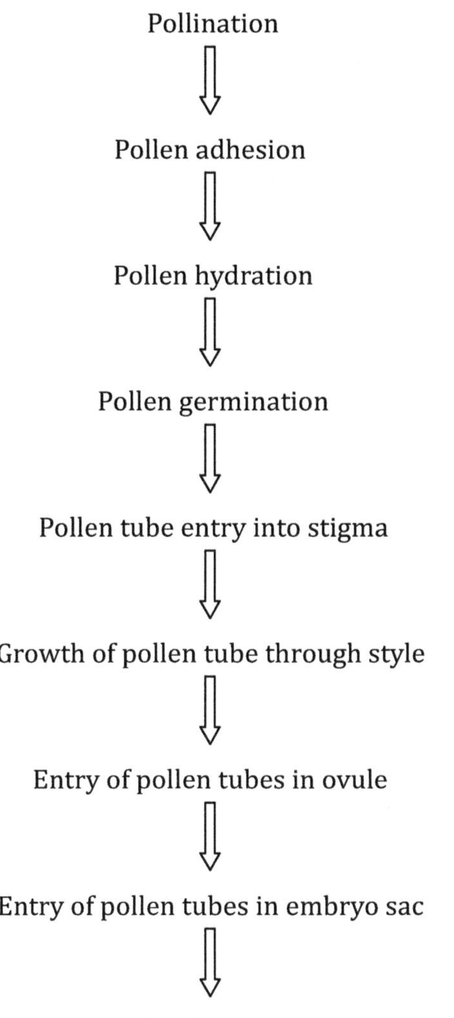

E. Sterility

Koelreuter in 1763 first reported male sterility in flowering plant. Later on numerous cases of male sterility in angiosperms were reported. Allard (1960) and Duvick (1966) presented a good account of male sterility in flowering plants.

"Male sterility refers to a condition in which pollen is either absent or non-functional in flowering plants. In other word male sterility is defined as an absence or non-function of pollen grain in plant or inability of plants to produce or release functional pollen grain. The use of male sterility in hybrid seed production has a great importance as it eliminates the process of mechanical emasculation.

Male sterility is characterized by non-functional pollen grain while female gametes are functional."

Main Features of Male Sterility

- ✰ It is an important outbreeding device which prevents autogamy and permits allogamy. In other words male sterile plants produce seed only on cross-pollination with the functional pollen of other plants.
- ✰ Presence of male sterility leads to heterozygosity in a species as it promotes out breeding and reduces homozygosity due to elimination of inbreeding.
- ✰ Male sterility results from the action of nuclear gene/cytoplasmic gene or both. Male sterility is caused due to pollen or anther abortion.
- ✰ Male sterility occurs in nature through spontaneous mutations as well as can be induced artificially by chemical or physical mutagens.
- ✰ Male sterility can be observed in all diploid species of crop plant both wild and cultivated if properly investigated. Male sterility has been reported in several crop plants. Male sterility has some similarities and some dis-similarities with self-incompatibility.

Types of Male Sterility

The male sterility is of five types:

1. Genetic male sterility
2. Cytoplasmic male sterility
3. Cytoplasmic genetic male sterility
4. Chemical induced male sterility
5. Transgenic male sterility

Difference between Self-Incompatibility and Male Sterility

Self-Incompatibility	*Male Sterility*
Similarities	
It is an important out breeding mechanism	Same as in SI
It is found in nature	Same as in SI
It prevents autogamy and promotes allogamy	Same as in SI
It is used for hybrid seed production	Same as in SI
Dis-similarities	
Pollen is functional	Pollen absent or non-functional
Due to morphological, genetical physiological and biological causes	Genetical, cytoplasm or both
It may be heteromorphic, homo-morphic, gametophytic or sporophytic	It may be genetical, cytoplasmic or cytoplasmic genetic
Artificial induction is difficult	Artificial induction is easy

1. Genetic Male Sterility (GMS) or Nuclear Male Sterility (NMS)

Like any other morphological traits, particularly mono and oligogenic, this type of male sterility occurs in plant due to mutation of fertility locus situated on chromosomes within the nucleus. In this case cytoplasm is not involved in bringing the sterility. Only nuclear gene is involved for genetic male sterility. There could be three possible genotypes for this locus and only one of them is male sterile.

Fertile (R-line) = RR (Growth level)

Fertilie (B-line) = Rr (Less growth level)

Sterile (A-line) = rr (Non growth level)

Sterility: In it the cross of sterile and fertilie lines gives rise to (A line x B line = 1 fertile: 1 sterile), *e.g.,* Watermelon, Muskmelon, Pumpkin, Tomato (*Ps*-2-gene), Sugarbeet and Chilli

Fertility: In it the cross of sterile and fertile lines gives rise to (A line x R line = all O fertile progenies).

Main Features of Genetic Male Sterility

- ✰ It is controlled by nuclear gene termed as genic or genetic male sterility.
- ✰ Mostly it is controlled by a single recessive gene (ms or s) with monogenic inheritance. But dominant gene male sterility is also found in safflower and duplicate in cauliflower.
- ✰ It is mainly consists of A line and B line where A lines are male sterile line (mm) while B line is male fertile line (Mm) and maintained by crossing both. A line is used as female parent in hybrid seed production. The crossing for maintenance of genetic male sterility gives rise to 50 per cent A line and 50 per cent B line.
- ✰ It is used for production of seed propagation and vegetative propagated crop plants.
- ✰ It is unstable and may become fertile at low temperature.
- ✰ The male sterility alleles may rise spontaneously or it can be induced artificially and is found in several crops, *e.g.,* Tomato, Lima bean and Pegion pea.
- ✰ A male sterile line may be maintained by crossing it with heterozygous male sterile plant and such mating produces 1:1 MS: MF plant.

	m	**m**
M	Mm	Mm
m	mm	mm

Merits

Generally male sterility is usually recessive and monogenic. Hybrid seed production and crossing are easy. It is used with both seed and vegetatively propagated species, requires less area, labour, to maintain A and B lines.

Demerits

It is less stable and becomes fertile at low temperature. If influenced by temperature, it is called thermosensitive and if influenced by day length, then it is called photosensitive.

2. *Cytoplasmic Male Sterility (CMS)*

It occurs due to mutation of mitochondria or due to some other cytoplasmic factors or gene or plasma gene outside the nucleus, resulting in the transformation of the fertile cytoplasm into a sterile one. Nuclear genes are not involved. At the most only two kinds of genotypes are possible, one of them is sterile (A line) and another fertile (B line). The fertile cytoplasm is denoted by (F-B line) and sterile cytoplasm is denoted (S-A line), *e.g.,* Chilli (Arka Meghana, Shweta and Harit), Onion (Arka Kirtiman and Lalima) and Carrot (Pussa Nayanjyoti and Vasuda).

Sterility Maintainance

f	x	F	= f
Sterile	x	Fertile	= All progenies become sterile as they inherit
(A line)		(B line)	sterile cytoplasm from sterile seed parents

Fertility Restoration: Since there is no third type of genotype which can act as R-line as such restoration of fertility is not feasible. This type male sterility is useful only in crops where seed is not desired end product but it suits horticultural crops as horticultural crops are vegetatively propagated.

Main Features of Cytoplasmic Male Sterility

- ✰ It is controlled by cytoplasmic gene or plasma gene. The F_1 seed (cross seed) produces only male sterile plants because their cytoplasm is derived entirely from the female gametes.
- ✰ This system consists of A line and B line. A is male sterile and B is male fertile and all other characters of A and B are similar or they are isogenic lines. No R lines are involved.
- ✰ This CMS line is maintained by crossing of A line with B line.
- ✰ It cannot be utilized for hybrid seed production without the use of restore line because F_1 seed produces male sterile F_1 plants.
- ✰ It can be used for development of hybrids in vegetatively propagated crop and ornamental crops where seed again is not the economic product.
- ✰ It is not influenced by environmental factors such as low or high temperature. In other word the sterility is stable.
- ✰ Usually the cytoplasm of zygote comes primarily from the egg cells and due to this progeny of such male sterile plant would be male sterile.

Merits

It is highly stable and is not influenced by environmental condition such as temperature and day length. Parental maintenance requires less area because breeder has to maintain A and B line only.

Demerits

It has limited application because it cannot be used for development of hybrid in those crops where seed is economic product. It can only be used in asexuall propagated species such as potato and forage crop. It is governed by plasma gene.

3. *Cytoplasmic Genetic Male Sterility (CGMS)*

When pollen sterility is controlled by both cytoplasmic and nuclear genes it is known as CGMS. Such sterility arises from the interaction of nuclear gene (s) and sterility with sterile cytoplasm. The cytoplasmic genetic sterility is essentially a cytoplasmic sterility with a provision for restoration of fertility. The fertility is restored by (R) gene present in nucleus. The combination of both nuclear gene (s) and cytoplasmic factors determines the fertility or sterility in such plants. Based on these combinations there can be six types of genotype and only one of them is sterile.

(i)	RR gene with F cytoplasm	RR F	(Fertile R-line)
(ii)	RR gene with f cytoplasm	RR f	(Fertile R-line)
(iii)	Rr gene with F cytoplasm	Rr F	(Fertile)
(iv)	Rr gene with f cytoplasm	Rr f	(Fertile)
(v)	rr gene with F cytoplasm	rr F	(Fertile B-line)
(vi)	rr gene with f cytoplasm	rr f	(Sterile A-line)

Sterility Maintainance

rr f	x	rr F	= rr f
Sterile (A-line)	x	Fertile (B-line)	= All male sterile progenies

Fertility Restoration

rr f	x	RR f or RR F	= Rr f
Sterile (A-line)	x	Fertile (R-line)	= Fertile hybrid

CGMS lines are of immense importance in exploitation of hybrid vigour in crops where seed is desired end product.

For Example

- Single recessive gene (ms), *e.g.*, Onion, Radish, Sweet pepper, Turnip (S-cytoplasm type utilized) and Cole crops (Ogura cytoplasm).
- Two or double recessive genes, *e.g.*, Beet
- Single dominant genes, *e.g.*, Carrot (Petaloids type)

Main Features of Cytoplasm Genetic Male Sterility

- The male sterility is controlled by the interaction of cytoplasm and nuclear genes (restoring fertility).
- This system includes A, B and R lines. A line is male sterile, B is similar to A in all features but is male fertile and R restores the fertility in F_1 hybrid and hence is called restorer. Since B line is used to maintain the fertility, it is also referred to as a maintainer line.
- CMS line can be maintained by crossing the mm x Mm or in other words maintained by crossing A and B lines.
- It is used for production of hybrid seed in seed propagated and vegetative propagated crops.
- This is highly stable and there is no effect of the temperature.

Merits

It is used in hybrid seed production in both sexual and asexual species. It is highly stable and not influenced by the effect of temperature.

Demerits

It requires more area and labour because breeder maintains three types of materials (A, B and R line).

4. *Chemical Induced Male Sterility*

Male sterility originates in nature as a result of spontaneous mutation. In some species male sterility has been derived from wild species through interspecific hybridization. It can also be induced through various chemicals. Main difference between mutagen induced male sterility and gametocide induced male sterility is that the former in heritable and the latter is non-heritable.

Advantage of Gametocide Induced Male Sterility

1. It is rapid methods of developing male sterility line. The backcross method takes 4-5 years for transfer of male sterility from one genotype to another. By the use of gametocide male sterility can be induced in one season.
2. This method is less laborious and less expensive than back cross method.
3. There is no need of maintaining A, B and R line. The genotype which has to be used as female parent in the development of hybrid can be treated with male gametocide to suppress pollen formation resulting in induction of male sterility, *e.g.,* Cucurbites (NAA), Sunflower, Onion, Lettuce (GA_3), Tomato, Cucurbits, Onion (MH) and Tomato, Sugarbeet (FW-450). However, this method has not be commercialized so far.

Drawback of Male Gametocides or Hormones

- Pollen abortion is incomplete and erratic.

- ☆ Used in specific stage and short lived effect again repeated treatment.
- ☆ Ovule fertility is also adversely affected which leads to low seed setting.
- ☆ Harmful on leaf and plant growth and very costly.

5. *Transgenic Male Sterility*

The foreign gene (transgene) is also used for induction of male sterility. The male sterility which is induced by the technique of genetic engineering is called transgenic male sterility. The gene responsible for inducing male sterility is generally used from micro-organism. This type of male sterility comes under genetic male sterility and is heritable. In this system two types of transgene are used.

One gene (barnase) causes male sterility hence this gene in integrated in A line. Another gene (barstar) suppresses the male sterility gene (barnase) and leads to restoration of fertility. When a cross is made between A and R line the F_1 is fertile. This is expensive method for commercial hybrid seed production of various crops.

GMS- msms x MsMs = Msms (male fertile)

Msms x Msms = Msms (male fertile 50 per cent)

MSms x MSms = msms (male sterile 50 per cent)

CMS- [Msms] S x [msms] F = S (male sterile)

CGMS- Msms (S) x msms (F) = (S) msms (male sterile 50 per cent) and (S) Msms (male fertile 50 per cent).

Transfer of male sterility: The male sterility can be easily transferred from one genotype to another genotype by back cross method. Generally 6-7 back crosses are sufficient to transfer male sterility into the background of a popular variety. For transfer of genetic male sterility the adapted variety is used as male parent and male sterile genotype as female parent. The F_1 is male fertile which segregates in 3 fertile: 1 sterile ratio in F_2 generation.

Sources of male sterility: Male sterile plants either do not produce pollen or produce non-functional or non-viable pollen. Male sterility plants have shriveled anthers and can be easily identified by visual observations of anthers. This is mainly of three types:

1. **Spontaneous mutation:** Both genic and cytoplasmic male sterility have been reported to occur by spontaneous (natural) mutation in various crop plants, *e.g.,* Sweet pea, Pea, Tomato, Brinjal and Lettuce.
2. **Induced mutation:** Male sterility can also be induced artificially by mutagens. Genic male sterility has been induced by X-rays in *Petunia*. In *Tagets errecta* an apetalous and antherless recessive mutant has been induced by Gamma rays. Genic male sterility has been induced by Gamma rays in tomato and watermelon and by chemical mutagens in pepper and pea.

3. **Inter-specific crosses:** Wild species of crop plants are important sources of sterile cytoplasm and sterile cytoplasm from wild species is transferred to cultivated species through interspecific hybridization and back crossing.

Utilizations: Male sterility has important applications in the development of hybrids in plant breeding. All the three types of male sterility are used in crop improvement programme.

Back cross: It refers to crossing F_1 with either of its parents, when F_1 crossed with homozygous recessive parent, it is known as test cross.

F. Dichogamy

The term Dichogamy is derived from the Greek word where '*Dikho*' means Apart and '*Gamous*' means Marriage. It refers to maturation of anthers and stigma of the same flowers at different times. Dichogamy promotes cross-pollination even in the hermaphrodite species. It is found mainly of two types.

1. **Protandry:** When anther (androcium) mature before pistil (gynoecium), the phenomenon is called as protandry. This is especially found in plants in which male reproductive organs mature before the female reproductive organs. Protandrous fruit crops are Walnut, Pecanut, Coconut, Annona, Aonla and Passion fruit and vegetable crops are Onion, Carrot, Muskmelon, Pointed gourd, Rhubarb, Parsnip, Leek, Garden beet and Swisschard
2. **Protogyny:** When pistil (gynoecium) matures before the anther (androecium), the phenomenon is called as protogyny. It is relating to a flower in which the shedding of pollen occurs after the stigma has stopped being receptive having female sex organs maturing before the male. In cherimoya flowers their stigmas usually lose their receptively before anther shed their pollen. Very commonly all flower on a single tree and even in on orchard synchronize their sexual cycle and most of the open flowers are either in female or in male phase with sparse overlap through the season. Due to this fact and the above mentioned absence of efficient pollinators in many countries, natural pollination commonly leads to low fruit set and small fruits of reduced marketability, *e.g.,* fruit crops are Banana, Fig, Annona, Pomegranate, Plum, Sapota, Strawberry, Avocado, *etc.* and vegetable crops are Chilli, Cole crop, Okra, Cassava, Amaranthus, *etc.*
3. **Heterodichogamy:** In differs from normal dichogamy in that it involves two matting types (protogyny and protandry) that occur at a 1:1 ratio in a population. Flowering phases of the two matting types are synchronized and reciprocal crosses occur to ensure between types out crossing. This study aims to quantify of flowering patterns and pollination efficiency in flower or plant in a wind pollinated heterodichogamy tree, *e.g.,* Pecanut, Pistachionut, *etc.*

4. **Protogynous diurnally synchronous dichogamy (PDSD):** The basic facts of avocado flower behaviour have been deciphered. It is protogynous diurnally synchronous dichogamy operates under normal condition in the flowering way, *e.g.*, Avocado.
   - ☆ Each and every flower opens twice. At the first opening it is functionally female the pistil is mature, the stigma is receptive while the pollen sacs remained closed. At the second opening the flower is functionally male dehiscence of the pollen sac occurs and pollen is being shed while upper part of the style is frequently shriveled and brown.
   - ☆ All flowers of the tree or cultivars are female before noon and male in the afternoon (A group) or are female in afternoon and male before noon (B group).
   - ☆ The elapsed time between the first and second opening is about 24 hr. for A group flower and about 12 hr. for B group flowers as reported by Bergh in 1975 in avocado.
5. **Duodichogamy:** Flowering plant commonly separate male and female function in time but rarely are the two stages synchronized within and among individuals. One such temporal matting system is duodichogamy in which each plant produces two batches of male flowers that are temporally separated by a batch of female flowers with or within individual synchrony and among individual a synchronous to ensure mating partners. Duodichogamy is known only in a few species in four genera in unrelated families, *e.g.*, in fruit crops is Chestnut, Chinese tree species Brideliatomentosa (Phyllanthaceae).

Effect and Uses of Dichogamy

It affects pollination and fertilization processes due to dichogamy resulting into unfruitfulness and irregular bearing if cross-pollination does not occur in absence pollinizers and pollinators. Because of dichogamy self-pollination is not possible so cross-pollination occurs or in some horticultural crops fruits develop or produce parthenocarpically. It is useful in hybridization of horticultural crops. If protandry and protogyny type phenomena are found we can improve those genotypes of the crops via cross-pollination. So dichogamy is good in that condition where we want to pollinate plant with oneself then this phenomenon is required. It has some disadvantage in case of farmers who usually do not known about these things so it problematic for them to produce crop because if they do not known about it (dichogamy type of flower behavior) and they do not do anything for artificial pollination. They won't use pollinators and pollinizers.

Chapter 5

Breeding Methods

Introduction

The production of new improved varieties of crops from local varieties that are inferior in many ways is called crop improvement. At present time more than 372 of vegetable, 72 of fruits, floriculture crops and 200 varieties of plantation crops are already released in India to enhance the crop production. However, there is a need for continuous crop improvement to resolve the problem of food shortage ahead.

Objectives of Crop Improvement

The main objectives of crop improvement are to produce improved varieties which are superior to their parent in all characters.

1. High yield potential.
2. Better quality of vegetable, fruit and flower crops.
3. Resistance to disease, insect and pest.
4. Resistance to abiotic stresses such as drought, frost, wind, flood, salinity and acidity.
5. Early maturation.
6. Better plant type.
7. Adaptability to changing environment.
8. Extended harvest period in vegetable and fruits.
9. Photosensitivity in vegetables.
10. Synchronus maturity in fruit and vegetable.

11. Better nutritional quality and processing quality.
12. Long shelf-life and good export quality in vegetable and fruits.

Methods of Crop Improvement

Crop improvement is the production of crop varieties with superior characters from inferior local varieties. In general crop improvement methods are divided into two categories.

1. Conventional method
2. Non-conventional method

(A) Conventional Method

The long established old ways of improving crops are called conventional methods or traditional methods. These are as follows:

a. Introduction
b. Selection
c. Hybridization
d. Ploidy breeding
e. Mutation breeding
f. Rootstock breeding

a. Introduction

The process of introducing plants from their native country to a new country is known as plant introduction. Plant introduction pertains to taking a genotype or a group of genotypes into a new environment where they were not being grown before. In other word plant introduction refers to transposition of crop plant from place of their cultivation to such area where they were never grown earlier. The systematic introduction and exchange of plant genetic resources of horticultural crops started in 1946 by division of botany IARI New Delhi which was replaced by the plant introduction and exploration organization in 1956 and then by the division of plant introduction in 1961. Later in 1976 National Bureau of Plant Genetic Resources (NBPGR) was established in New Delhi. It has been done when a variety growing in donor country is superior to indigenous varieties develop in the recipient country. Plant introduction involves two basic process namely introduction and acclimatization. Under introduction plants are transferred from the native country to another country having a different climate. The adaptation of introduced plants to survive in new area is called acclimatization. It is determined by mode of pollination, genetic variability in individuals, longevity of crop and susceptibility of mutations.

The important cultivars and species of different horticultural crops introduced from time to time for various purposes are as under:

- ☆ **Banana:** Cultivar Lady Finger (EC 160160) from Australia as resistant or tolerant to bunchy top diseases and Grand Naine MS (EC 27237) from France and West Indies.
- ☆ **Citrus:** Torocco Blood from USA, Sunraman from Peru, Vainigilia Sanguigno (pigmented) from USA and Tha and Thai-1 from Thailand.
- ☆ **Date palm:** Tayar, Hatemi, Mejnaj, Khalas, Ruziz and Khesab from Arabia, Samani Zaghlool, Hayani, Samy, Ari, Agolani, Saklote and Amhat from Egypt.
- ☆ **Fig:** Genoa White, 0030 Geneo, 0009 Flandes, Candria (Adriatic type fig) from USA, Mission, Ficus, Rocdin-3 and Ficus Standard from USA.
- ☆ **Grape:** Thompson Seedless, Perlette and Beuty Seedless from USA, Kishmish Beli and Kismish Charmi from USSR, Totlocha from Brazil and Dogridge from USA.
- ☆ **Mango:** Tomy, Ziulets, Haden, Sensation, Julie from USA.
- ☆ **Guava:** Beaumont G-35 and Indonesia Seedless from Australia, Verdie and M-25988 from USA.
- ☆ **Papaya:** PI-41436, Soniyimma, Malinchalu from Nigeria, Sunrise, Wilder, Sunset T88/211, *Carica cauliflora* from USA.
- ☆ **Pomegranate:** Wonderfull from USA, Rannij G-1-8-23 and Rannij G-1-3-34 from USA.

The main points related to plant introduction are briefly discussed-

1. This is an oldest method of crop improvement. It is used in all three group of crops *viz.*, self-pollinated, cross-pollinated and asexually propagated. In other word this species is applicable to all three types of crop species.
2. Plant introduction is usually done from one country to other country. But sometimes it may take place between climatic regions of the same country.
3. Earlier the introduction of plant from one place to another place used to be done by travelers, traders, invaders and merchants. Now specific organization has been established for this work (see later).
4. The plant material is obtained through expeditions, personal visit and correspondence, *e.g.,* Change of material, purchase or as a gift.
5. The material of seed propagated crop is introduced in the form of seed and that of vegetatively propagated crops in the form of cutting or propagules.

Source of Introduction

There are five important source of plant collection:

- **a.** Centres of diversity
- **b.** Gene bank

- **c.** Gene sanctuaries
- **d.** Seed companies
- **e.** Farmers field

Types of Plant Introduction

It is generally classified on the basis of adaptation and utilization. Based on utilization introduction are of two types:

1. Primary introduction.
2. Secondary introduction.

Based on adaptation introduction are of two types.

1. Direct introduction
2. Indirect introduction

1. *Primary Introduction*

Introduction that can be used for commercial cultivation as a variety without any change in the original genotype is referred to as primary introduction. Introduction that are immediately adapted to the changed environment is known as direct introductions. Thus primary introduction can also be called direct introduction. Any foreign variety which is directly recommended for commercial cultivation in the new environment (country) is called exotic variety.

2. *Secondary Introduction*

Introduction that can be used as a variety after selection from the original genotype or used for transfer of some desirable gene to the cultivated variety is known as secondary introduction. Secondary introduction is more common than primary introduction those introduction that take same years for adaptation to the new environment are termed as indirect introductions. Thus secondary introduction can also be called as indirect introduction.

Procedure of Introduction

Procedure of germplasm In the form of gift, exchange, purchase or collected through an exploration quarantine (to keep materials in isolation to parent the spread of disease, *etc.*)

Cataloguing EC- for exotic collection, IC-for indigenous collection

IW- for indigenous wild collection

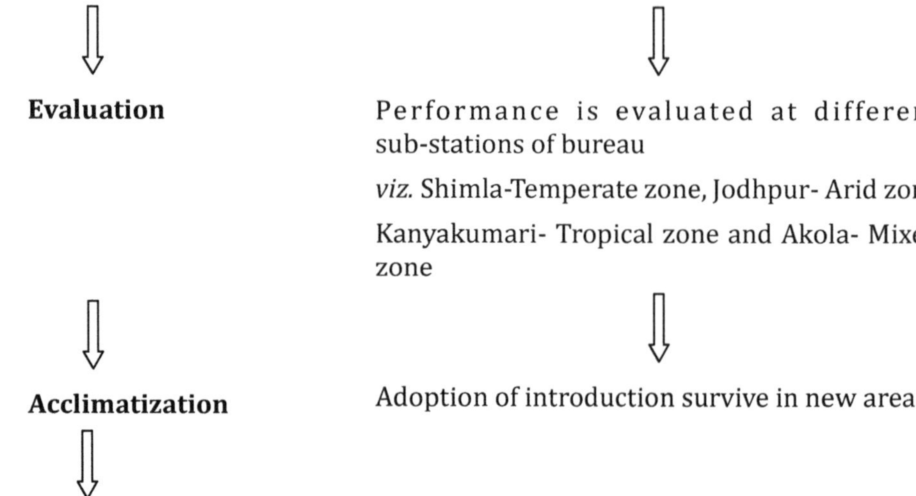

Multiplication and distribution

Merits of Introduction

1. It is very easy and quick method of developing new variety especially when introduced material is used directly as a variety or after selection
2. This method is useful in introducing new crops plants. Crops like potato, tomato, papaya, pineapple, *etc.* were introduced in India from other countries.
3. This is good method of collection and conservation of germplasm of different crop to protect the same from genetic erosion.
4. This is an effective method of conserving those crop species which are in the danger of extinction. Such species can be saved by shifting them to other areas.
5. This method is applicable in all self-pollinated, cross-pollinated and vegetative propagated crops.
6. Plant may be introduced in new disease free area to protect them from damage.
7. It is quick and economical method of crop improvement.

Demerits of Introduction

The main demerits of this method are that there are chances of the entry of new diseases, insect and weeds in the country along with the introduced material. For example, *Argimone maxicana* weed, late blight of potato, bunchy top of banana and coffee rust diseases and wooly aphids of apple and potato tubers moth's insects were introduced in India along with introduced material. However, all these entered before the establishment of quarantine organization. Now chances of entry of new insect, disease and weeds are remote due to quarantine regulation and check up.

b. Selection

This is an art of choosing better individuals from a population to raise a new variety. It is a conventional methods and oldest method of plant breeding. Nature itself constantly selects fittest individuals to survive and destroys all other things slowly. Therefore, all local varieties are the product of natural selection. The selection of better individual from existing population in the crop field is called artificial selection. Selection is the basic method in almost all method of plant breeding. There are three selection methods, for crop improvement namely mass selection, pure line selection and clonal selection. Most of the present-day cultivars of horticultural crops (fruit crops) are the result of open-pollination seedlings selection from the mixed population for the selection there must be.

Large population, variation must be present in the population and variation must be heritable.

Achievements

Name of fruit crops: Name of the cultivars

- ✰ **Mango:** Langra, SB Chausa, Sukul, Dashehari, Bombay Green, *etc.*
- ✰ **Guava:** L-49, Arka Mridula, Allahabad Safeda, Pant Prabhat and Lalit.
- ✰ **Litchi:** Swarn Roopa.
- ✰ **Aonla:** Kanchan (NA-4), Krishana (NA-5), (NA- 6) and (NA-7).
- ✰ **Papaya:** CO-1, CO-2, CO-3, CO-5, CO-6, Pusa Delicius, Pusa Majesty, Pusa Giant, Pusa Dwarf and Coorg Honey Dew.
- ✰ **Kagazi lime:** Vikram, Paramalini, Sai-sarbati, Jaidevi and Pant Lemon.
- ✰ **Bael:** NB-5, NB-9, CISHB-1, CISHB-2, Pant Arparna, Pant Shivani, Pant Urvashi and Pant Sujata.
- ✰ **Sapota:** CO-2 and PKM-1.

(i) Mass Selection

Mass selection refers to a method of crop improvement in which individual plants are selected on the basis of phenotype from a mixed or heterozygous population; their seeds are bulked and used to grow the next generation. This is one of the oldest methods of crop improvement. Mass selection is applicable to both self and cross-pollinated species. However, it is more commonly used in the improvement of cross-pollinated crops (onion and cabbage) than in self-pollinated species. This method is rarely used in vegetativelly propagated crops.

Main Features of Mass Selection

1. **Genetic Constitution:** In self-pollinated crops a mass selected variety is homozygous but heterozygous, because it a mixture of several purelines. In cross-pollinated crop such varieties are mixture of homo

and heterozygotes and are heterogynous, because they consist of several homozygous and heterozygous genotypes.

2. **Adaptation:** Mass selection varieties have wide adaptation and are more stable against environmental changes due to heterogeneity which provides better buffering capacity. However, adaptability is more in cross-pollinated crops than in self-pollinated species.
3. **Variation:** Mass selection varieties are composed of several purelines in self-pollinated species crops and of several homozygous and heterozygous genotypes in cross-pollinated crops. Hence there is heritable variation in the mass selection varieties besides environmental variation.
4. **Selection:** It is effective in case of mass selection varieties of self-pollinated crops due to presence of heritable variation however further selection in the mass selection varieties of cross-pollinated crops may lead to inbreeding depression.
5. **Quality:** A variety developed by mass selection is less uniform in the quality of seed than purelines due to presence of heritable variation.
6. **Resistance:** Mass selection varieties are less prone to the attack of new diseases due to genetic diversity. In other word they are more resistance or tolerant to new diseases.
7. **Roguing:** Periodic removal of off type plants is essential to maintain the yield of mass selection varieties.

Types of Mass Selection

It is found mainly of two types:

1. **Positive mass selection:** When desirable plants are selected from a mixed population and their seeds are mixed together to grow further generation, it is referred to as positive mass selection. This process is continued for several years.
2. **Negative mass selection:** When only undesirable off type plants are removed from the fields and rest are allowed to grow further, it is known as negative mass selection. This is generally used for varietal purification in seed production and certification programmes.

Procedures of Mass Selection (1-8 years)

First year- An old variety or landrace is grown in large plot then individual plants (55-100) are selected as phenotypic base. The selected plants are harvested at maturity and their seeds are mixed together and grow in next generation.

Second year- Crop is grown from bulk seed of selected plants in separate field using standard variety as check and materials are evaluated in preliminary yield trial. The check variety is used for comparison.

Third to sixth year- The performance of bulk is evaluated for yield and adaptation in main yield trials from 3-4 years using standards check for comparison.

Seventh to eight year- Variety is released following trials in all India co-ordinated research project and due process of release and named in VIIth year and seed is multiplied in VIIIth year and seed is ready for distribution.

Merits of Mass Selection

- ✰ This is a good method for improvement of old varieties and landraces. This is also used for the purification of improvement variety.
- ✰ Mass selection varieties are more stable in their performance than purelines. In other words, they have more buffering capacity than purelines due to heterogeneities.
- ✰ Mass selection is a simple and quick method of crop improvement. It takes about 8 years for the release of a new variety whereas, pureline selection takes about 10 years in the development of new variety.
- ✰ This method is applicable to both self and cross-pollinated species.
- ✰ Mass selection varieties provide good protection against diseases.

Demerits of Mass Selection

- ✰ This selection is based on the phenotypic performance. The superior phenotype is not always an indication of superior genotype. The real breeding value of single plants can be judged from the performance of their progeny. Progeny test is not carried out in mass selection.
- ✰ In cross-pollinated species, there is no control on the pollination. The selected plants are pollinated by both superior and inferior pollen of parents. This results in rapid deterioration of variety developed by mass selection.
- ✰ In cross-pollinated crop large numbers of plants have to be selected for bulking because small sample will lead to inbreeding depression.
- ✰ The produce of varieties developed by mass selection is less uniform than pure line. This is because mass selection varieties are mixture of several purelines in self-pollinated crops and consist of several genotypes in cross-pollinated species.
- ✰ In self-pollinated species pureline selection is more effective than mass selection because pureline selection leads to isolation of the best line from a mixed or heterogeneous population.

Progeny Selection

A selection procedure in which superior plants are selected from a heterogeneous population on the basis of their progeny performance is referred to as progeny selection. Progeny selection refers to selection of plants from a diverse population on the basis of their progeny test. The test of genotypic value of an individual based on the performance of its progeny is called progeny test. The progeny test was developed by Louis de Vilmorin hence it is also known as Vilmorin principle. It is used for:

- Understanding whether a plant is homo and heterozygous.
- In the assessment of the breeding value of a plant.

(ii) Pure Line Selection

The concept of pureline selection was developed in middle of 19th century in Sweden (Le-Couteur, Shireff, Hallet and Vilmorin between 1840-1860). However, the genetic basis of pureline was explained by W.L. Johanson, a Danish biologist in 1903. Pure line refers to the homogeneous progeny of a self-pollinated homozygous plant. Development of new variety through identification and isolation of single best plant progeny is known as pureline selection or individual plant selection. This method is commonly used in self-pollinated species.

Main features of pure line selection: Pure line selection is practiced in heterogeneous population such as introduced materials, landraces and mass selection varieties of self-pollinated species to isolate superior genotype.

1. **Homogenous:** All the genotypes of a pureline are homogeneous (homozygous), *i.e.,* genetically identical and phenotypically similar.
2. **Non-heritable variation:**-The variation within a pureline is entirely due to environmental factors. Thus the variation is non-heritable in the purelines.
3. **Highly uniform:** A variety developed by pureline selection is highly uniform for various traits due to absence of genetic variation.
4. **Selection within pure line ineffective:** Selection is ineffective in a pure line due to lack of heritable variation. Selection is effective when heritable variation is present.
5. **Narrow adaptation:** Generally pure line varieties have narrow adaptation and poor adaptability than heterogeneous population. The poor adaptability is due to narrow genetic base.
6. **Isolation of pure line:** Pure liners can be isolated from heterogeneous population as well as segregating population through individual plant selection and progeny selection.
7. **Source of variation:** In pure line variety, natural outcrossing, mutations and mechanical mixture are the important sources of genetic variation. The spontaneous mutation cannot be controlled other factors can be controlled.
8. **More prone to new disease:** Pure line varieties are more prone to the attack of new disease due to genetic uniformity and narrow genetic base.

Procedures of Pure Line Selection (1-10 years)

First year- An old variety or landrace is used as a base population, single plants are selected from the heterogeneous population keeping in view the objective of selection. The number of individuals plant selected may vary from 200-1000 in various crops.

Second year- Progeny of each selected plant is grown in few rows (progeny row) and evaluated for desirable characters. The top 15-20 progenies are selected and seed from plant in each selected progeny is bulked which constitutes strains.

Third year- The strains constituted in second year are evaluated in replicated field trials and top performing few strains are selected for further evaluation.

Fourth to seventh year- In India the selected entirely (strains) are evaluated in all India co-ordinated crop improvement project and best genotype is selected on the basis of yield performance.

Eight to tenth year- The best performing strains is released and notified as a variety. Then the breeder, foundation and certified seeds are produced. The produce of certified seed takes two years after release of a variety. Thus the seed of new variety reaches the farmers in tenth year.

Merits of Pure Line Selection

- This is good method of isolating the best genotype for yield, disease resistance, insect resistance, earliness, quality, *etc.* from a heterogeneous or mixed population of an old variety.
- The variety developed by this method is uniform and more attractive than mass selection variety.
- This is an easy and cheap method of crop improvement.

Demerits of Pure Line Selection

- This method can isolate only superior genotype from the mixed population. It can not develop new genotype.
- This method is applicable to self-pollinated species only. It can not be used for development of variety in cross-pollinated species.
- The varieties developed by pure line selection have poor adaptability due to narrow genetic base. All the plants of a pure line have identical genotype. Hence such varieties are more prone to the attack of new disease due to genetic uniformity.

(iii) Clonal Selection

Progeny of a single plant obtained by asexual reproduction is known as clone. A procedure of selecting superior clones from the mixed population of asexually propagating crop is referred to as clonal selection and selected clones are propagated asexually or vegetative means. There are some horticultural crops (banana, mango, citrus, apple, pear, peach, loquat, litchi, *etc.*) that are propagated by asexual means.

"A clone is a group of plants produced exclusively from a single individual plant through asexual propagation or asexual reproduction. Most of the fruit crops are propagated asexually which consist of large number of clones that is why these plant are known as a group of plant derived from a single plant by vegetative means. In other words all the vegetative progenies of a single plant make a clone".

The main reasons of asexual reproduction are as follows:

- ✰ Reducing flowering and seed set time.
- ✰ Non-flowering in many cases.
- ✰ To avoid in breeding depression in certain crops.
- ✰ Apomixis in some species.

Non-flowering species- This group includes garlic, ginger, betel and several yams.

Low seed setting species- Potato and sweet potato.

Normal flowering and seed setting species- Citrus, mango, pear, peach, apple, litchi, loquat and many ornamental plants. These are highly heterozygous and vegetative propagation is essential to maintain the heterozygous balance.

Characteristics of Clone

1. **Homogeneous constitution:** The progeny of a clone is genetically identical thus clone are homozygous. In other word individual plant of a clone is a mitotic derivative of the same plant and therefore homogeneity in phenotype is major feature of clone. A group of individual plant derived from the same tissue of the original mother plant carries the same genotype, phenotypic variation if any in clones is due to environmental impact.
2. **Heterozygosity:** The asexually propagated crops are heterozygous and hence clone is also heterozygous. Progeny of a clone looks similar phenotypically but is heterozygous. If a clone is subjected to inbreeding, it will produce various types of segregates and exhibit high inbreeding depression. In other word continuous inbreeding of clones which are heterozygous might lead to severe loss in vigor.
3. **Vigorous growth:** Clones have hybrid vigour which is conserved due to asexual reproduction, most of potato varieties are hybrids. In other word clonal selection is useful in conserving the heterosis for a long period because clones are stable (they retain their original traits just like pureline variety) and are not prone to segregation.

4. **Wider adaptability:** Generally clones are more adaptable to environmental variation due to high level of heterozygosity than pure line. A delibrate mixture of genetically different but phenotypically similar clones gives better yield in variable environments than a single clone. Theoretically clones are immortal, *i.e.,* a clone can be maintained indefinitely by asexual reproduction. However, these are very much susceptible to disease or insect pest depending upon the species and cultivars.
5. **Source of variation:** There are three sources of variation in a clone *viz.,* bud mutation, mechanical mixtures and occasional sexual reproduction. The frequency of bud mutation is very low but once bud mutation occurs it will lead to deterioration of a clone by adding new variants in the population. Viral and bacterial diseases also lead to deterioration of a clonal variety.
6. **Segregation in F_1:** When hybridization is done between different clones, segregation occurs in F_1 generation. Each F_1 plant is potentially a new variety; therefore selection is practiced in F_1.
7. The phenotype of a clone is due to effect of gene (G), environment (E) and GxE interaction. Therefore P=G+E+GE.
8. Clones are maintained by asexual reproduction but purelines and inbred are maintained by self-pollination or close inbreeding.
9. **Clonal degeneration:** The loss in vigour and productivity of clones with the passing of time is known as clonal degeneration and it may be due to mutation, infection of virus and bacteria
   a. **Mutation:** Somatic mutations are also known as bud mutations. The frequency of mutations is generally very slow. A mutant allele would be homozygous only when (i) both the alleles in the cell mutate at the same time producing the same mutant allele or (ii) the mutant allele is already in the heterozygous condition is the original clone. Bud mutations often produce chimeras, *i.e.,* individuals containing cells of two or more genotypes. However, it is not a great problem because normal plant, *i.e.,* non-chimeras may be produced from chimeras by several techniques.
   b. **Mechanical mixture:** Mechanical mixture produce genetic variation within a clone similar to the manner as seen in purelines.
   c. **Sexual reproduction:** Occasional sexual reproduction leads to segregation and recombination. The seedlings obtained from sexual reproduction are genotypically different from the asexual progeny.

Procedures of Clonal Selection (1-9 years)

The phenotypic value of a plant or a clone is due to its genotype (G), the environment (E) and the genotype x environment interaction (GxE) of these only the

G effects are heritable and stable. Therefore a selection for quantitative characters based on single plant observation may not hold good.

A selection for polygenic characters like yield on the basis of unreplicated clonal plots trial would also often be misleading and unreliable. The value of clone can be reliably estimated only through replicated yield trials. However, selection for highly heritable characters such as plant height, days to flowering, colour, disease resistance, *etc.* is easy and effective even on the basis of single plant or plot. The various steps involved in clonal selection are briefly described below and are depicted.

First year: From a mixed variable population, a few hundred to few thousand desirable plants are selected. A rigid selection can be done for simply inherited characters with high heritability plants with obvious weakness are eliminated. In fruit plant it is difficult to get large number of individual selection. In such case, few plants may be selected.

Second year: Clones from the selected plants are grown separately, generally without replication. This is because of the limitation in propagation materials in each clone and also because of the large number of clones involved. The

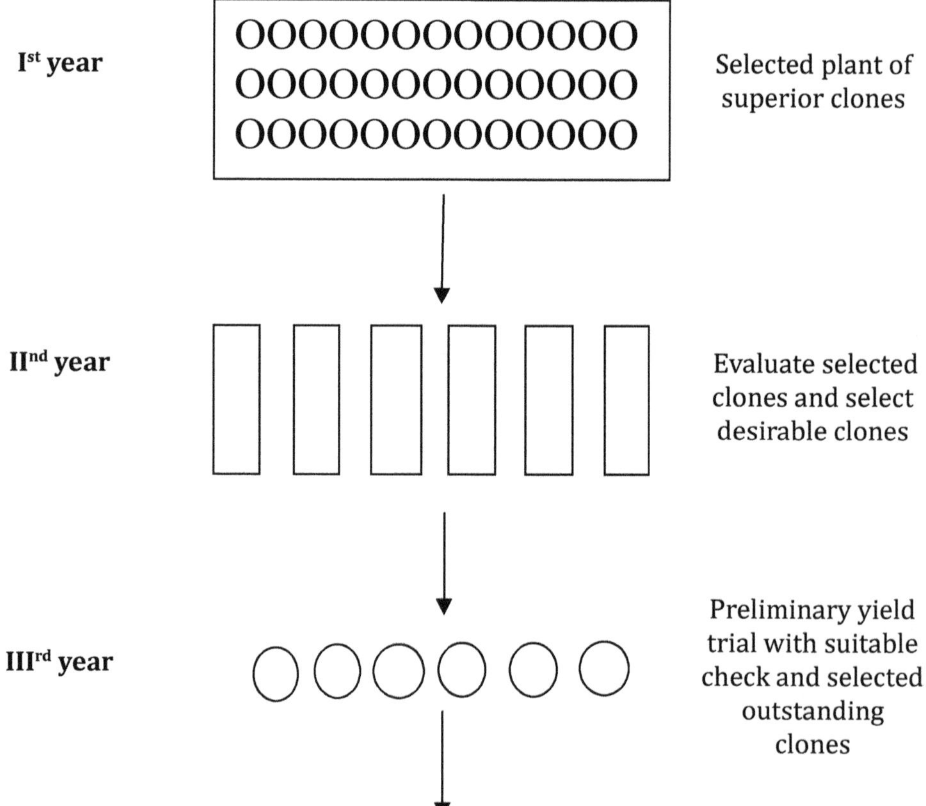

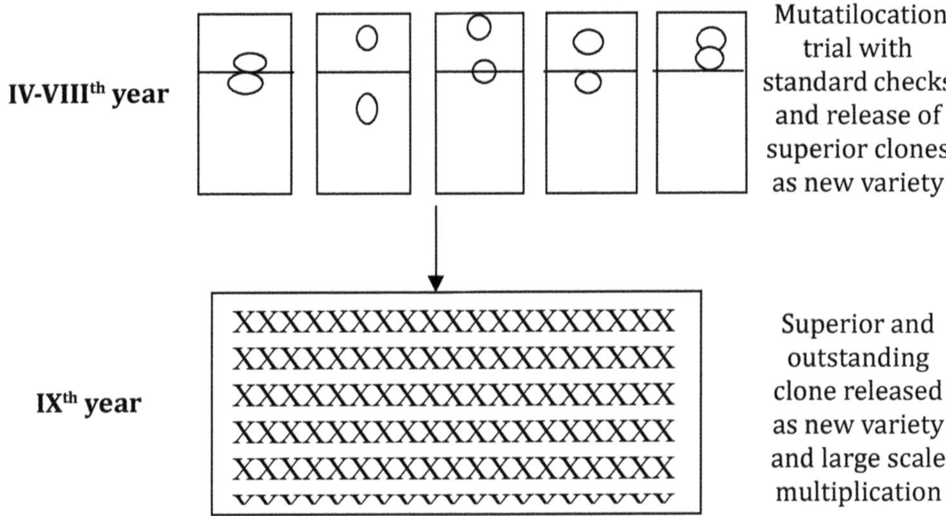

characteristics of clones will be clear now than in the previous generation when the observation were based on single plant. The inferior clones are eliminated at this stage the selection is based on visual observation and on the breeder's judgment of the value of clones. Fifty to one hundred clones are selected on the basis of clonal characteristics.

Third year: Replicated preliminary yield trial is conducted and a suitable check is included for comparison. Few superior performing clones with desirable characteristics are selected for multilocation trials. At this stage selection for quality is done if necessary. Separate disease nurseries may be planted to evaluate disease resistance of the selected clones.

Fourth to seventh year: Replicated yield trials are conducted at several locations along with a suitable check. The yielding ability quality and diseases resistance, *etc.* of the clones are rigidly evaluated. The best clones that are superior to the check in one or more characteristics are identified for release as variety.

Ninth year: The superior clones are multiplied and released as varieties.

Advantage of Clonal Selection

1. This method is useful in conserving heterosis for several generations. The variety evolved by this method retains all the characteristics of the parental clone for several years.
2. This method is easy for maintenance and less time consuming because there is no problem of outcrossing and loss of seed viability.
3. Heterotic clones on selection may be used as permanent hybrids. Heterosis can be exploited for longer time without production of hybrid seed every year (for vegetatively propagated vegetable crops).

4. This is an effective method for genetic improvement of vegetatively propagated crop plants. In other word it is useful in isolating the best genotype from a mixed population of asexually propagated crops.

Disadvantage of Clonal Selection

1. Varieties developed by clonal selection are highly prone to new races of a disease.
2. Clonal selection cannot create new variability and therefore genetic makeup cannot be improved by this method without hybridization.

Limitations

- There is limited chance of getting new and useful type of variability.
- The multiplication rate is low.
- It is only useful for vegetatively propagated plants.

Achievements: Clone No.-51 from Dasehari, MA-1 from Alphonso, Tommy Atkins from Haden, and Pusa Surya from Elden in Mango and Pusa Seedless from Thompson Seedless in Grape.

Some Important Varieties of Horticultural Crops

- **Mango:** Chinnaswarnrekha (bud grafting) and Niranjan from Parbhani.
- **Grape:** Tas-A-Ganesh, Cheema Sahiba, Rao Sahiba and Sarad Seedless (from Kishmish Charni).
- **Banana:** Batheesas, Pidi Monthon and High Gate (from Grass Michel).
- **Potato:** Kufari Red and Kufari Safed (from Phulwa).
- **Colocasia:** Sree Rashmi and Sree Pallavi (from Trivendrum Local).
- **Rose:** Edward Rose and Andhra Rose.
- **Marigold:** Orange and Yellow.

Vegetative propagule of horticultural crop plant by clonal selection

- **Stem cutting:** Grape, sweet potato, drumstick, betel vine, pepper and ornamental plant.
- **Suckers:** Banana, pineapple, aloe and agave.
- **Bulb:** Onion and garlic.
- **Rhizome:** Ginger.
- **Graft and buds:** Mango, citrus, apple and ornamental plant.

c. Hybridization

The crossing of two genetically different parents is called hybridization. The resulting offspring are called hybrids. In other word the mating or crossing of two plants or lines of dissimilar genotypes are known as hybridization. It creates variability among the individuals of the progeny due to recombination. This variability is inherited to offspring for several generations. Hybridization was first used in plant improvement by Joseph Kolreuter in 1760 but this method became popular after Mendel's discovery on the plant hybridization. Artificial hybridization is performed in both self and cross pollinated crops.

Objectives of Hybridization

- To bring together desired characters of different plants in one variety.
- To bring out hybrid vigour in individuals.
- To achieve genetic variability in plants due to genetic recombination.
- To get plants to improve crops via selection method.
- To restore vigour lost during continuous inbreeding of crops.

Types of Hybridization

It is of following types:

1. Inter-varietal Hybridization

The parents involved in hybridization belongs to the same species. They may be two strains, varieties or races of the same species. It is also known as intraspecific hybridization. The inter-varietal crosses may be simple or complex depending upon the number of parents involved.

a. Simple Cross

In a simple cross two parents are crossed to produce the F_1

b. Complex Cross

More than two parents are crossed to produce the hybrid.

Three parent crosses (A, B, C)

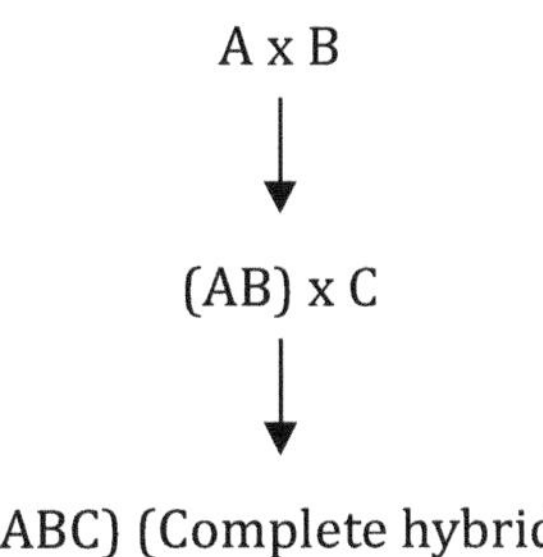

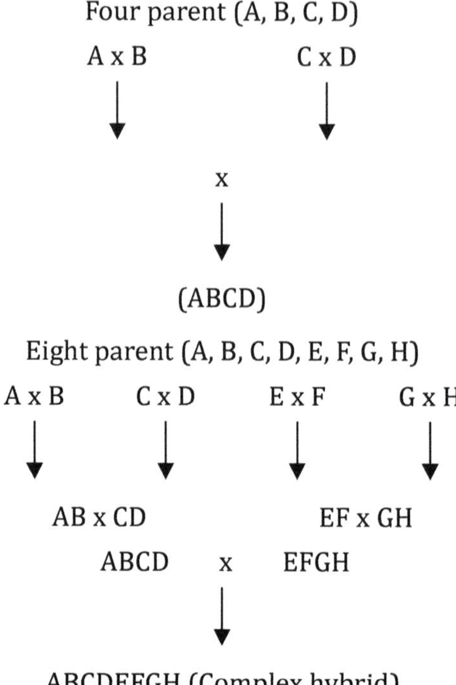

2. Distant Hybridization

Distant hybridization includes crosses between different species of the same genus or of different genera. When two species of the same genus are crossed it is known as inter-specific hybridization but when they belong to two different genera it is inter-generic hybridization. In other word crossing between two different species of the same genus or two different genera of the same family is called distant hybridization and such crosses are referred to as distant crosses or wide crosses. Wide crossing or distant hybridization has been used in the genetic improvement of some crop plants. Distant hybridization is of two types:

- **a.** Inter-specific hybridization.
- **b.** Inter-generic hybridization.

a. Inter-specific Hybridization

Crossing or mating between different species of the same genus is referred to as inter-specific hybridization. Because inter-specific hybridization involves two species of the same genus it is also termed as intera-generic hybridization.

Main features of inter-specific hybridization

- ✰ It is used when the desirable character is not found in the species of a crop.

- ☆ It is an effective method of transferring desirable genes in to cultivated plants from their related cultivated or wild species.
- ☆ It is more successful in vegetatively propagated species like potato than in seed propagated species.
- ☆ Interspecific hybridization gives rise to three type of crosses *viz.,* fully fertile, partially fertile and fully sterile in different crop species.
- ☆ Fully fertile and partially fertile crosses are useful in agronomic crops and fully sterile crosses in horticultural crop.

Fully sterile crosses: Inter-specific crosses are fully sterile between those species which do not have chromosomal homology. In such species chromosome number may or may not be similar. The lack of chromosomal homology does not permit pairing between the chromosomes of two species during meiosis. As a result the F_1 plants are fully self-sterile such hybrid can be made self-fertile by doubling of chromosome through colchicines treatment, *e.g., Brassica spp.* Inter-specific crosses in genus Brassica were made by several workers. Three crosses were made among species manly cabbage (*Brassica oleracea*), rapeseed (*Brassica campestris*) and black mustard (*Brassica nigra*). The F_1 hybrids were sterile in all the three crosses.

Crosses		F_1	*New Species*	*Remarks*
Brassica nigra x *Brassica oleracea*		Sterile colchicine	*Brassica carinata*	Fertile
2n=16	2n=18	treatment 2n=17	2n=34	BBCC
(BB)	(CC)	(BC)		
Brassica nigra x *Brassica campestris*		Sterile, colchicine	*Brassica juncea*	Fertile
2n=16	2n=20	treatment 2n=16	2n=32	AABB
(BB)	(AA)	(AB)		
Brassica oleracea x *Brassica campestris*		Sterile, colchicine	*Brassica napus*	Fertile
2n=18	2n=20	treatment 2n=19	2n=38	AACC
(CC)	(AA)	(AC)		

b. *Inter-generic Hybridization*

Inter-generic hybridization refers to crossing between two different genera of the same family. Such crossing are rarely used in crop improvement because of various problem associated with them.

Main features of inter-generic hybridization

- ☆ It is used when the desirable genes are not found in different species of the same genus.
- ☆ It has been generally used in asexually propagated species.
- ☆ F_1 hybrid between two genera is always sterile. The fertility has to be restored by doubling of chromosome through colchicines treatment.

1. **Radish and cabbage cross:** Inter-generic crosses between radish (*Raphanus sativus*) and Cabbage (*Brassica oleracea*) of the family Cruciferae was made by Karpenchenko in 1928 in Russia. The main objective was to combine root of radish with leaves of cabbage. The F_1 was sterile. The doubling of chromosome number by colchicines treatment resulted in development of fertile amphidiploids which were named as Raphanobrassica by Karpenchenko. But the new species thus develop hads root like cabbage and leaves like radish which was an useless combination.
2. **Tobacco and tomato cross:** Inter-generic cross between tobacco (*Necotiana tobacum*) and tomato (*Solanum lycopersicum* L.) of the family Solanaceae.
3. **Potato and tomato cross:** Inter-generic cross between potato (*Solanum tuberasum*) and tomato (*Solanum lycopersicum* L.) of the family Solanaceae.
4. **Intertribal hybrids:** Hybrid between plants of two different tribes is called intertribal hybrids such hybrid can only be produced via somatic hybridization. Such crosses exhibit more asymmetric hybrids then inter specific and inter generic hybrids. Based on combination of chromosomes and cytoplasm from different parental species somatic hybrids are of three types:
   - **Symmetrical hybrids:** Those somatic hybrids cells or plants that contain all chromosomes of both the species involved in the fusion are referred to as symmetrical hybrid, *e.g.,* Carrot and *Petunia.*
   - **Asymmetrical hybrids:** Those somatic hybrid cells or plant which contain complete somatic complement of one species and only a part of somatic complement of another species are called asymmetrical hybrids, *e.g.,* Potato, Petunia and inter generic and intertribal hybrids.
   - **Cybrid:** Those somatic hybrid that involve normal protoplast of one species and nucleus protoplast (cytoplast) of another species are known as cybrid, *e.g.,* Transfer of cytoplasmic male sterility from one species to other.

Procedures of Hybridization Programme

The technical operation of hybridization work is known as hybridization technique. There are seven steps involved in hybridization.

1. **Selection of parents:** The parent plants is selected from the population of individual growing in isolated crop field in the farmers land or in regional research station. In general parent plant are selected from closely related varieties or species, but in some cases distantly related plants are used

as parent plants usually one parent is chosen from a well established variety having one weak character and the other parent is chosen from another variety that can complement the weakness. In mainly depends upon the objective of breeding programme. In addition to other objectives, increased yield is always the objective of breeder.

2. **Evaluation of parents:** If the performance of parents used in breeding is known an evaluation is not necessary. But if their performance is not known, they should be evaluated particularly for the characters to which they are expected to contribute.
3. **Emasculation:** The removal of stamens or anther from the flowers of female parent before the dehiscence of anthers or the killing of pollen grains of a flower without affecting in any way to female reproductive organs is known as emasculation. It is done a few hours before the opening of flower buds. The purpose of emasculation is to prevent self-fertilization in the flowers of female parents. One day before opening of flower, bud is selected for emasculation.

   Types of Emasculation

   ✰ **Hand emasculation:** In species with large flowers, removal of anthers is possible with the help of forceps or scissors. The base of flower is held between thumb and index finger of the left hand, with the right hand the flower bud is opened using forceps and then stamen are pulled out with the forceps, *e.g.,* Tomato, Brinjal, Bottle gourd, *etc.*

   ✰ **Suction emasculation:** It is useful in species with small flowers. A thin rubber or a glass tube attached to a suction hose is used to suck the anthers from the flowers. The amount of suction used is very important which should be sufficient to suck the pollen and anthers but not gynoecium.

   ✰ **Hot water emasculation:** In plants with very small sized flowers emasculation is done by hot water method, *e.g.,* the panicle is kept dipped in hot water (at 45-53°C) in a jug for 1-10 minutes. The hot water inactivates the stamen to avoid self-pollination.

   ✰ **Alcohol or PGR treatment:** Treatment with alcohol, 2,4-D or NAA or Malic hydrozide during the growth stage. It is better method of emasculation than suction method.

   ✰ **Genetic emasculation:** Male sterility is introduced into plants by backcross method. Then the male sterile plants are used as a female parent and then there is no need for emasculation. This method is often called male sterility method.
4. **Bagging:** Protecting emasculated flowers or panicles from contaminating pollen grains by enclosing in a butter paper bag or muslin bag of

appropriate size to prevent random cross-pollination is called bagging. The paper bags immersed in oil or wax protect flower from insect attack, in most cases bagging is done in the evening. The bag size is determined by the size of flower of flower heads or panicle. The bags are tied at the base using a thread or copper wire. The male and female flowers are bagged separately. Bags are removed from the male plants after collecting the pollen grains from the stamens on the other hand bags are retained on female flowers and panicles up to the stage of seed development.

5. **Labelling or tagging:** Emasculated flowers are tagged just after bagging. The following information is recorded on the tag with a carbon pencil.
   - Number allotted to the field and plant.
   - Date and time of emasculation.
   - Date and time of pollination.
   - Name of female and male parents. The name of female parent writing first and that of male parent writing latter.
6. **Crossing or cross-pollination:** The transfer of pollen grains collected from male plant to emasculated female flowers is called crossing or artificial cross-pollination. It involves the collection of pollen from desired male parent and dusting them on the pistil of emasculated flower. The pollen grains are collected from dehisced anther of male flowers in a petri dish or in paper bags. The stamens are gently teased with a forceps or brush to release the pollen from them then released pollens are collected in a petri dish. Stigma of emasculated flower is receptive in the morning hours so that crossing is usually done in the morning. In case of pollination, mature, fertile and viable pollen should be placed on a receptive stigma to bring fertilization.
7. **Harvesting and storage of F_1 seed:** After reaching maturity the bags are removed from the plant and crossed heads or panicles are harvested from the plants. Hybrid seed are collected from the harvested heads dried and stored separately in polythene bags. In other word crossed fruits along with seed should be harvested. For recalcitrant seeds sowing is done immediately after extraction of seed. However, orthodox seed of some horticultural crop plant can be stored for longer time. Hybrid seed obtained from the hybridization are sown separately in the field to raise F_1 generation in the next season. All F_1 plants are heterozygous but have similar genetic composition. These plants are said to be a hybrid variety. They may or may not show hybrid vigour.

Hybridization Techniques

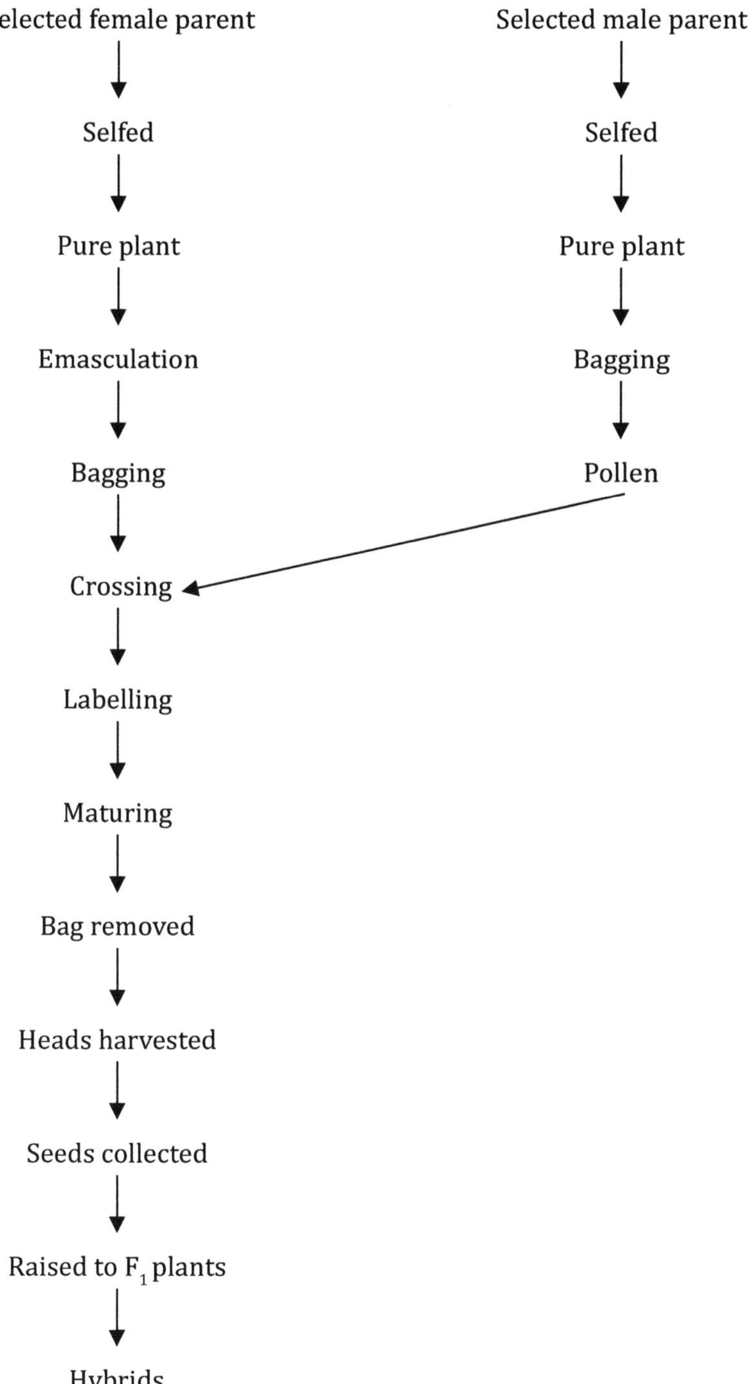

Hybridization Method

In order to produce improved varieties, F_1 hybrid is selfed or crossed with yet other parent or one of the parents in the previous cross. This method of hybridization based crop improvement is called as hybridization method or selfing technique.

1. Inter-specific Hybridization

Hybridization in which one species is crossed with yet other species belonging to the same genus is called inter-specific hybridization. This is also known as intra-generic hybridization. It is performed to transfer of disease resistance, pest tolerance and drought resistance from species to another species, *e.g.*

- ✰ The potato variety Kufri Kuber is obtained from a cross between *Solanum curtilobum* and *Solanum andigenum.* Kufri Sindhuri is obtained from cross between *Solanum andigenum* and *Solanum tuberosum.*
- ✰ The tomato variety Pusa Red Plum is the product of inter specific hybridization between *Lycopersicum esculentum* and *Lycopersicum pimpinellifolium.*

2. Intra-specific Hybridization

Hybridization in which one variety is crossed with another variety belonging to the same species is called intra-specific hybridization. This is also called intra-varietal hybridization.

- **A.** Pedigree method
- **B.** Bulk method
- **C.** Backcross method
- **D.** Single cross method
- **E.** Double cross method
- **F.** Multiple cross method

A. Pedigree Method

The word pedigree refers to a line of ancestors. Selection of an improved variety from a line of ancestors established from F_1 individuals is known as pedigree method. It is widely used by plant breeders to improve self-pollinated crops. It is a quick method of crop improvement in releasing new variety in 9-10 years. This method is best suited to improve visible phenotypic characters in plants. Here progenies of superior plants only are selected to proceed the next generation.

Pedigree method of plant breeding

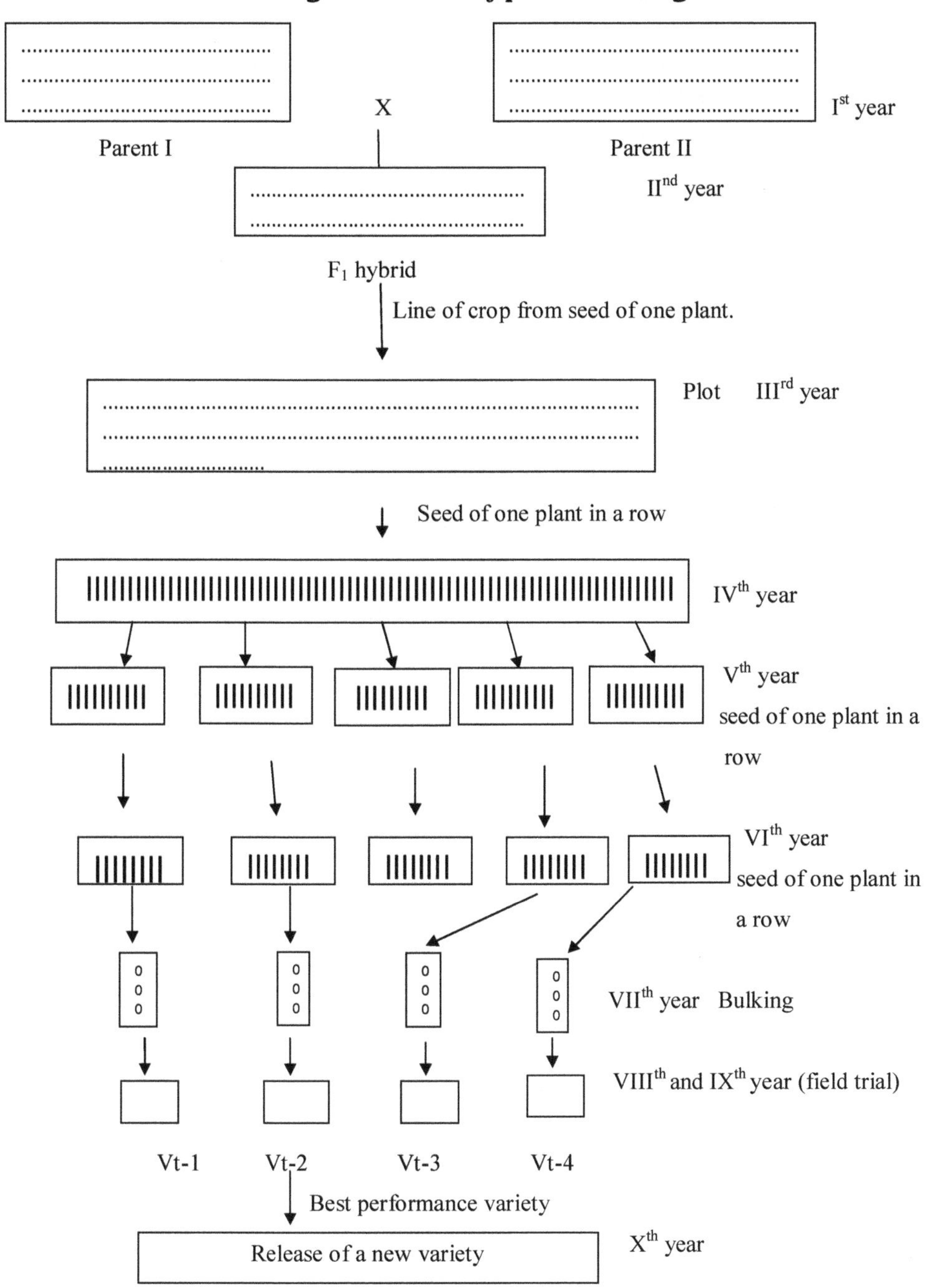

B. Bulk Method

Breeding in which seeds of superior individuals are collected bulked and sown repeatedly until a new variety is formed is called bulk method. It is used in breeding self-pollinated crops.

Bulk method in plant breeding

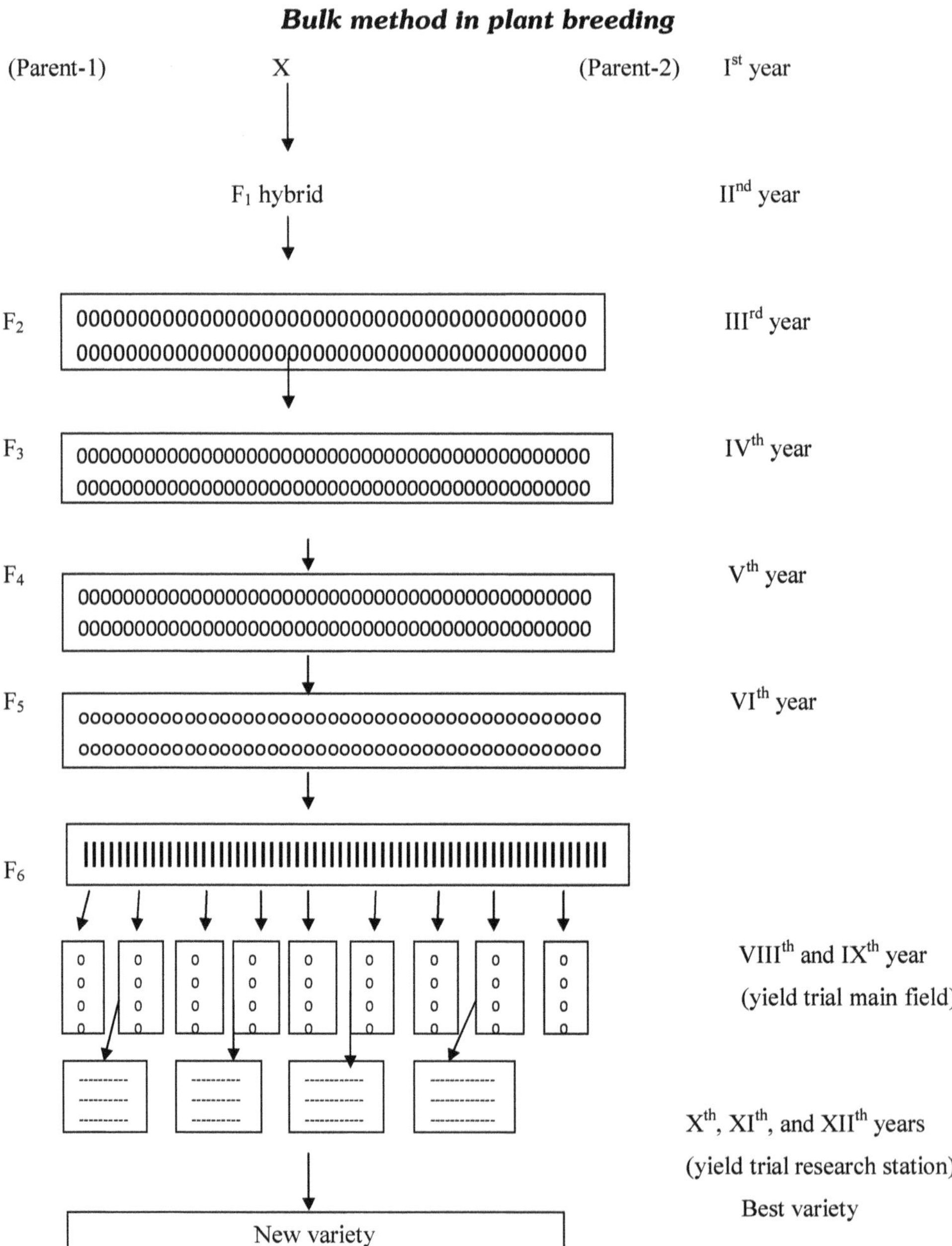

C. *Backcross Method*

Hybridization method in which the F_1 hybrid is crossed with one of its parents is called as backcross method. It was first used in plant improvement by Harlan and Pope in 1922. When one or two characters have to be transferred to otherwise superior variety, this is most commonly used method. Backcross is a recurrent hybridization, *i.e.,* the cross is repeated once again with a parent. The parent plant with one or two superior characters is called recurrent parent or recipient and inferior parent is called donor parent.

D. *Single Cross Method*

In this method hybrids are obtained from two inbred parents to get improved hybrids. This method was proposed by Shull in 1909. It helps to get hybrid vigour in multiple allelic inheritances.

Improvements of hybrid by single cross method

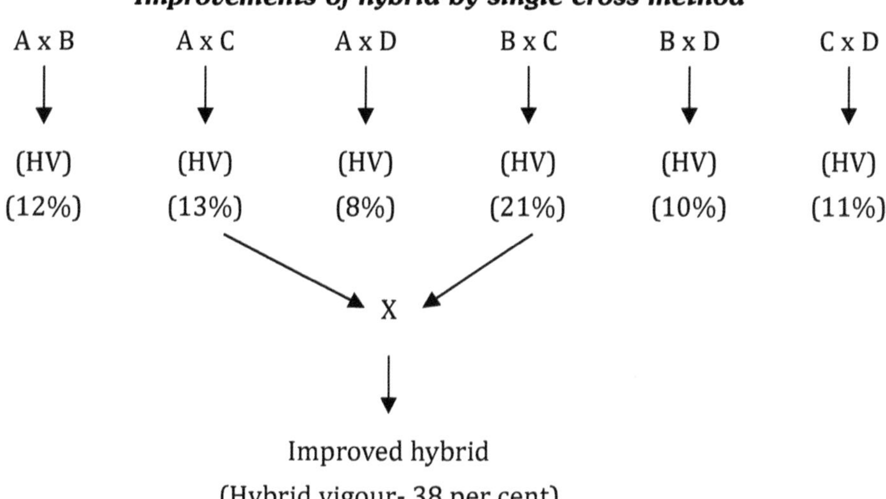

E. *Double Cross Method*

In this method purelines for dihybrids are planted in alternative rows and crossed to get inbreed

Improvements of hybrid by double cross method

AAbb x aaBB → AaBb

CCdd x ccDD → CcDd

AaBb X CcDd

↓

Aa Bb Cc Dd (Improved hybrid)

F. Multiple Cross Method

This method is useful to transfer genes located in different chromosome for getting improve hybrid.

Improvements of hybrid by multiple cross method

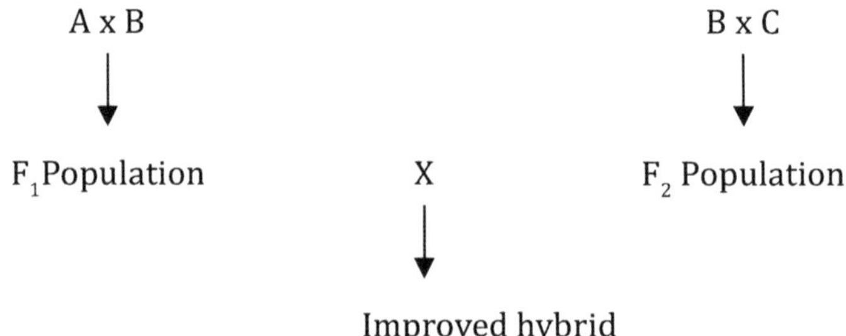

Achievement through Hybridization

(a) Vegetables

- **Tomato:** Pusa Ruby (Sioux x Improved Meruti), Punjab Chhuhara (Punjab Tropics x EC-55055), Hisar Lalima (Pusa Early Dwarf x H-5101).
- **Brinjal:** Hisar Jamun (Aushey x BR-34), Arka Navneet (IIHR 22-1 x Supreme), Pusa Kranti ([Pusa Purple Long x Hyderpur] x Wyand Giant).
- **Capsicum:** Green Gold and Pusa Deepti.
- **Chilli:** Andhra Jyoti (G-2 x Bihar variety), Pusa Jwala (NP-46-A x Puri red), Pant C-1 (NP-46-A x Kandhari, a natural cross) and K2 (K1x Sattur samba).
- **Okra:** Pusa Sawani (IC-1542 x Pusa Makhmali), Punjab Padmini [Rashmi x Ghana] x [Pusa Sawani x Ghana].
- **Cauliflower:** Pusa Snowball (EC-12012 x EC-12013), Pusa Snowball K1 (Pusa Snowball x Dutch Introduction).
- **Cabbage:** Pusa Mukta (EC-24855 x EC-10109).
- **Radish:** Pusa Himani (Black Radish x Japanese White), Pusa Rashmi (Green Type x Desi Type), Punjab Safed (White-5 x Japanese White).
- **Turnip:** Pusa Kanchan (Local Red Round x Golden Ball), Pusa Chandrima (Snowball x Japanese White) and Pusa Swarnima (Golden Ball x Japanese White).
- **Carrot:** Pusa Kesar (Local Red x Nantes), Pusa Meghali (Nantes x Pusa Kesar), Imperatar (Nantes x Chanteny).
- **Musk melon**: MHY3 (Durgapur Madhu x Pusa Madhuras), Punjab Rasilla (Pusa Madhuras x Durgapur Madhu).
- **Bitter gourd:** Phule Green Gold (Green Long x Delhi Local).
- **Bottle gourd:** Punjab Komal, Pusa Meghdoot and Pusa Sandesh.

- **Ridg gourd:** Arka Sujat and Surekha.
- **Pumpkin:** Pusa Hybrid-1.
- **Cluster bean:** Pusa Naubahar (Pusa Mausami x Pusa Sadabahar).
- **Sweet potato:** H41 (Norin x Local type), H42 (Kelladumph x Triumph).
- **Topioca:** H97 (Manjuvella x Brazillian), H165 (CG x Kilikalan), H226 (EK x M4), Sree Visakham (AC no.1501 x S2312), Sree Sahaya.

(b) Spices

- **Black pepper:** Panniyur-1 (Uthrinkota x Cherikaniakadam).

(c) Fruits

- **Mango:** M1 (Amrapali x Janardhan Pasand), Sindhu (Ratna x Alphanso), Ratna (Neelum x Alphanso), Mallika (Neelum x Dashehari), Amarapali (Dashehari x Neelum), Arka Neelkiran (Neelum x Alphanso), Arka Anmol (Alphanso x Janardan Pasand), Arka Puneet (Banganpalli x Alphanso), PKM1 (Chinnaswarnarekha x Neelum), Manjira (Rumani x Neelum).
- **Grape:** Arkavati (Black Champa x Thampson Seedless), Arka Kanchan (Anab-e-Shahi x Queen of Vineyard), Arka Neelamani (Black Champa x Thampson Seedless), Arka Majestic (Anab-e-Shahi x Black Champa), Pusa Urvashi (Hur x Beauty Seedless) and Pusa Navrang (Madeleine Angavine x Ruby Red).
- **Guava:** Safed Jam (Allahabad Safed x Kohir), Kohir Safed (Kohir x Allahabad Safed), Arka Amulya (Allahabad Safed x Seedless), Hybrid-16-1 (Apple Colour x Allahabad Safed).
- **Apple:** Lal Ambri (Red Delicious x Ambri), Ambstarking (Starking Delicous x Ambri 81), Sunehri (Ambri x Golden Delicious), Chaubattia Anupam (Delicious x Early Shanburry), Chaubattia Princess (Delicious x Early Shanburry).
- **Sweet Orange:** Calamondin (*Citrus reticulata* x Fortunella), Citrumelo (Trifolite orange x Grape fruit).
- **Mandarin:** Kinnow (King x Willow leaf).
- **Papaya:** CO-3 (CO-2 x Sun Rise Solo), CO-4 (CO-1 x Washington).
- **Pomegranate:** Mridula, Ruby.
- **Sapota:** CO-1 (Cricket ball x Oval), DHS-1, DHS-2, PKM-2, PKM-3.
- **Avocado:** Furete (Calinson x Winslowson).
- **Annona:** African Pride (Cherimola x Custard Apple), Arka Sahan (*Annona atemoya* x *Annona squamosa*).
- **Almond:** Sloh (Peach x Almond hybrid).
- **Strawberry:** *Fragaria annanasa* (*Fragaria chilonensis* x *Fragaria virginiana*).

- ☆ **Plum:** Santa Rosa [*Prunus salicina* x (*Prunus salicina* x *Prunus americana*)].
- ☆ **Peach:** Sahranpur Prabhat (Sarbati x Flordasun)

(d) Plantation Crops

- ☆ **Coconut:** VHC-1 (East Coast Tall x Malaysia Dwarf Greed), VHC-2 (East Coast Tall x Malaysia Dwarf Greed), VHC-3.
- ☆ **Cashewnut:** Vengurla-3 (Vengurla-1 x Vector 56), Vengurla-4 (Midnapore Red x Vector 56), Vengurla-5 (Ansur Early x Mysore Kotekan), Vengurla-7, Bapatla-1, Madakkanthara and Dhana.
- ☆ **Walnut:** Paradax (*Jaglans hindsi* x *Jaglans rugra*).
- ☆ **Coffee:** Selection No.-7 (San x Roman), Selection No.-6 (*Coffea congensis* x Robusta).

d. Ploidy Breeding

An organism or individual having more than two basic or monoploidy sets of chromosome is called polyploid such condition is known as polyploidy. It is estimated that about one third species of flowering plant are polyploids. In wild species of grass family polyploidy has been reported up to 70 per cent. In other word the individuals carrying chromosome number other than diploid (2x and not 2n) number are known as polyploid and the situation is known as heteroploidy. Polyploidy breeding seems to have prospects of obtaining large sized fruit with dwarf plant type.

"An organism having more than two sets of homologous chromosome (diploid organism or pair during meiosis) is known as a polyploid. Polyploidy is of general occurrence in plants while it is rare in amongst animals if represented by 'n' then the simple polyploidy series would be":

1. n- Haploid.
2. 2n- Diploid.
3. 3n- Triploid.
4. 4n- Tetraploid.
5. 5n- Pentaploid.
6. 6n- Hexaploid.
7. 7n- Heptaploid.
8. 8n- Octaploid.
9. 9n- Nonaploid.
10. 10n- Decaploid and so on.

Polyploidy is pervasive in plants and some estimates suggest that 30-80 per cent of living plant species are polyploids and many lineage show evidence of ancient polyploidy (Paleopolyploidy) in their genomes. Polyploids plant can arise spontaneously in nature by several mechanisms including meiotic or mitotic features and fusion of unreduced (2n) gametes.

Chromosome

In the nucleus of each cell DNA molecule is packaged into thread like structure. Each chromosome is made up of DNA tightly coined many times around proteins called histone that support its structure.

Behaviour of Polyploidy Crops

Polyploids plants tend to be larger and better at thriving in early succession habitats such as farm fields. In the breeding of crops the tallest and best thriving plants are selected. For example, many seedless fruit varieties are seedless as a result of polyploidy. Such crops are propagated using asexual techniques such as grafting polyploidy in crop plants is most commonly induced by treating seeds with the chemical colchicines.

Example of Polyploidy Crops

- **Triploid crops**- Banana, Apple, Ginger, Citrus, *etc.*
- **Tetraploid crops-** Potato, Leek, Peanut, Kinnow, Pelargonium, *etc.*
- **Hexaploid crops-** Chrysanthemum, Kiwifruit, *etc.*
- **Octaploid crops-** Strawberry, Dahlia, Pansies, *etc.*

Apple, Tulips and Lillies are commonly found as both diploid and triploid. Bananas are available as diploid, triploid tetraploid. Daylilies (*Hemerocallis* spp.) cultures are available as either diploid or tetraploid. Kinnow can be tetraploid, diploid or triploid.

Inducing of Polyploidy

Polyploidy is mainly induced by treatment with a chemical known as colchicine. This is an alkaloid which is obtained from the seeds of a plant known as *Colchicum autumnale*, which belongs to the family Liliaceae. It is applied in very low concentration (0.01 per cent to 0.5 per cent) in different plant species because the chemical is highly toxic.

Effect of Polyploidy

The distinct features of autoploids are increase in general vigour and size of various plant parts. Such features are generally referred to as gigantism.

1. Stem are thick and stouter.
2. Leaves are fleshy, thicker, larger and deep green in colour.
3. Root is stronger and longer.
4. Flowers pollens and seed are larger than diploids.
5. Maturity durations longer and growth rate is slow than diploids.
6. Water contents are higher than diploids, *etc.*

Types of Polyploidy

Polyploidy is mainly two types:

1. Autopolyploidy
2. Allopolyploidy

1. Autopolyploidy

Polyploidy which originate by multiplication of chromosome of a single species are known as autopolyploids or autoploids and such situation is referred to as autopolyploidy. In other word autopolyploids (self or small) are polyploids with multiple chromosome sets derived from a single species. They can result from a spontaneous, naturally occurring genome doublings like the potato. Others might arise from following fusion of 2n gamets (unreduced gametes). Banana and apple can be found as autopolyploids. Autoploids include triploids (3x), tetraploids (4x), pentaploids (5x), hexaploids (6x), septaploids (7x) octaploid (8x) and also so on.

(a) Autotriploids

They have three sets of chromosome of the same species. They can occur naturally or can be produced artificially by crossing between autotetraploid and diploid species. Triploids are generally highly sterile due to defective gamete formation.

- ☆ **Banana:** Cultivated varieties of banana are triploids and seedless. Such banana has larger fruit than diploid ones.
- ☆ **Apple:** One variety of apple is triploids which is propagated asexually by budding or grafting.
- ☆ **Sugarbeet:** Triploid sugarbeets have higher sugar content than diploids and are generally resistant to moulds.
- ☆ **Watermelon:** Triploids watermelon are seedless or have rudimentary seed like cucumber. They are produced by crossing of tetraploid female and diploid male.

(b) Autotetraploids

They have four copies of the genome of same species. They may arise spontaneously or can be induced artificially by doubling the chromosomes of a diploid species with colchicine treatment, *e.g.,* Grape- Triploid grapes have been developed in California, USA, which have larger fruits and fewer seeds per fruit than diploids.

2. Allopolyploidy

A polyploid organism which originated by combining complete chromosome sets from two or more species is known as allopolyploid or alloploid and such condition is referred to as allopolyploidy. Alloploids are also known as hybrid polyploids or bispecies or multispecies polyploids. In other words allopolyploids are polyploids with chromosome derived from different species. Precisely it is result of doubling of chromosome number in an F_1 hybrid. Amphidiploid is another word for an allopolyploid. Mango and banana are also allopolyploids, doubled diploids are known as amphidiploids, *e.g.*, Brassicas- Three diploid Brassicas (*Brassica oleracea*, *Brassica rapa* and *Brassica nigra*) and three allotetraploids (*Brassica napus*, *Brassica carinata* and *Brassica juncea*) have been reported.

e. Mutation breeding

Mutation refers to sudden heritable change in the phenotype of an individual. In the molecular term, mutation is defined as the permanent and relatively rare change in the number or sequence of nucleotide. In other words, mutations arise due to change in DNA bases. The change in the nucleotide may involve replacement of one purine by another pure (A ⟷ G) or one pyrimidine by another pyrimidine (C ⟷ T). Such change is called transition. Thus mutations occurs in two main ways:

- ✰ By alteration in nuclear DNA (point mutations).
- ✰ By change in cytoplasmic DNA (cytoplasmic mutation).

Types of Mutations

Mutations do occur in nature (spontaneous mutation) as well as can be artificially induced by various mutagenic agents (induced mutation). Induced mutations are of two types:

1. **Macro-mutations:** Mutations with distinct morphological change in the phenotype are referred to as macro-mutations. Such mutations are found for qualitative characters and therefore are also called oligogenic mutations. Its identification is very easy.
2. **Micro-mutations:** Mutations with invisible phenotypic changes are called micro-mutations. Such mutations are observed in quantitative characters and hence are also referred to as polygenic mutations. Their identification is very difficult.

Mutation Breeding

The genetic improvement of crop plants for various economic characters through the use of induced mutations is referred to as mutation breeding. It is one of the special methods of crop improvement. Mutation breeding is commonly used in self-pollinated and asexually propagated species however, this method is rarely used for genetic improvement of crops-pollinated species.

Mutagens and their Mode of Actions

Mutagens refers to various physical or chemical agents which greatly enhance the frequency of mutations. Mutagens are mainly of two types:

A. Physical Mutagens

Physical mutagens include various types of radiation, *viz.,* X-rays, Gamma rays, alpha rays, beta rays, particles fasts and thermal (slow) neutrons and ultra-violet rays.

Particulate: Alpha rays, beta rays, fast neutron, thermal neutron.

Non particulate: X- rays and Gamma rays.

a. Ionizing Radiations

1. **X-rays:** X-rays were first discovered by Roentgen in 1895. The wavelengths of X-rays vary from 10^{-11} to 10^{-7}. They are sparsely ionizing and highly penetrating and are generated in X-rays machines, X-rays can break chromosomes and produce all type of mutations in nucleotides. X-rays were first used by Muller in 1927 for induction of mutation in *Drosophila*. In plants stadler in 1928 first used X-rays for inductions of mutation in barley.
2. **Gamma rays:** Gamma rays have shorter wavelength and are more penetrating than X-rays. They are generated from radioactive decay of some elements like 14°C, 60°C radium, *etc.*, of this cobalt 60 is commonly used for the production of Gamma rays.
3. **Alpha particles:** It is composed of alpha particles. They are made of two protons and two neutrons and thus have double positive charge. They are densely ionizing but lesser penetrating than beta rays and neutrons.
4. **Beta particles:** Beta rays composed of beta particles. They are sparsely ionizing but more penetrating than alpha rays.

b. Non-ionizing

Ultra-violet radiation.

B. Chemical Mutagens

1. **Alkylating agents:** The alkylating agents inched, Ethyl Methane Sulphonate (EMS), Methyl Methane Sulphonate (MMS) Ethylene Imines (EI) Sulphur mustard, Nitrogen mustard, *etc.*
2. **Base analogous:** It refers to chemical compounds which are very similar to DNA bases. Such chemical sometimes are incorporated in DNA in place of normal base during replication (5-bromouracil and 5- chlorouracil).
3. **Acridine dyes:** It is very effective mutagen. It includes proflavin acridine orange, acridine yellow and acriflavin and ethidium bromide.
4. **Other mutagens:** Other chemicals are nitrous acid and hydroxylamine.

Process of Mutation Breeding

The mutation breeding, treating a biological material with a mutagen in order to induce mutation is known as mutagenesis. When mutations are induced for crop improvement the entire operation of induction and isolation of mutants is termed as mutation breeding. The mutation breeding prosses consist of mainly four important steps and other various steps involved in mutation breeding.

- ☆ **Choice of variety**: Best important variety choices for mutation breeding.
- ☆ **Choice of mutagen:** Many mutagen are choice for the mutation breeding.
- ☆ **Mutagenic treatment:** Three things are taken into account, *viz.*, plant species, dose of mutagen, duration of treatment. In seed propagated species generally seeds are treated with dose (LD-50).
- ☆ **Handling of treated material:** It is differs in seed propagated and vegetatively propagated crops. In seed propagation mainly five steps. Steps I. M_1 generation, II. M_2 generation III. M_3 generation, IV. M_4 generation, and V. M_5- M_9 generation are used.

In vegetative propagation five steps- Steps I. VM_1 generation, II. VM_2 generation, III. VM_3 generation, IV. VM_4 generation, and V. VM_5- VM_9 generation.

The various steps involved in mutation breeding are as under:

- ☆ **Objective of programme:** Objective should be clear cut and well defined.
- ☆ **Selection of variety:** Locally accepted best variety in which improvement is needed either in polygenic or monogenic trait.
- ☆ **Part of plant to be treated**: Seed pollen grain or vegetative propagule (buds and cutting) may be used for mutagenesis. Selection of plant parts of mutagenic treatment is based on mode of multiplication or reproduction. In sexually propagated fruit plants, seed treatment is common. Pollen grain may be used but it has some limitations such as it is difficult to collect large amount of pollen grains and pollen survival life is short. In case of asexually propagated fruit plants buds or cutting are used for mutagenic treatment.
- ☆ **Dose of mutagen:** An optimum dose is that one which produces the maximum frequency of mutation and causes the minimum killing, *i.e.*, LD-50. It is the dose of mutagen which would kill 50 per cent of treated individual dose of mutagen depends upon intensity and time of treatment.
- ☆ **Mutagen treatment:** Selected plant part is exposed to the desired mutagen dose. In the case of irradiation the plant are immediately planted to raise M_1 plants from them. In the case of seed treatment they are pre-soaked for a few hours to initiate metabolic activities and then exposed to mutagen. Treated seeds are sown immediately in the field to raise M_1 generation when pollen grain are treated germination resulting from the seeds produced by treated pollen grain would be the M_1, M_2, M_3, M_4, generation, *etc.* are subsequent generations, derived from M_1, M_2, M_3, *etc.* plant through selfing or clonal propagation.

General Characteristics of Mutation

1. Mutations are generally recessive but dominant mutations also occur.
2. Mutations are generally harmful to the organism.
3. Mutations are random.
4. Mutations are recurrent.
5. Induced mutations commonly show pleiotropy often due to mutations are closely linked to gene(s).

Advantage of Mutation Breeding

1. Induced mutagenesis is used for the induction of cytoplasmic male sterility.
2. Ethidium bromide (EB) has been used for the induction of cytoplasmic male sterility in pearl millet and barley.
3. Mutation breeding is a cheap and rapid method of developing new varieties as compared to backcross, pedigree and bulk breeding method.
4. Mutation breeding is more effective for the improvement of oligogenic characters, such as disease resistance than polygenic traits.
5. Mutation breeding is the simple quick and best way when a new character is to be induced in vegetatively propagated crops.

Disadvantage of Mutation Breeding

1. Most of the mutations are deleterious and undesirable.
2. Identification of micro-mutations which are more useful to a plant breeder is usually very difficult.
3. Mutation breeding has limited scope for the genetic improvement of quantitative or polygenic characters.

Procedures of Mutation Breeding

Selection of plant materials seed, seedling and cutting

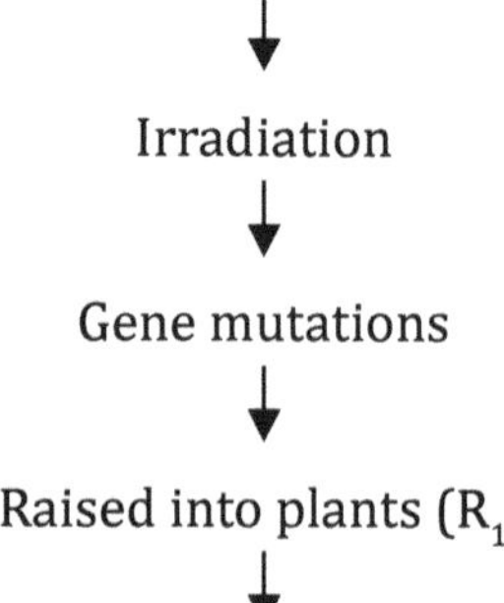

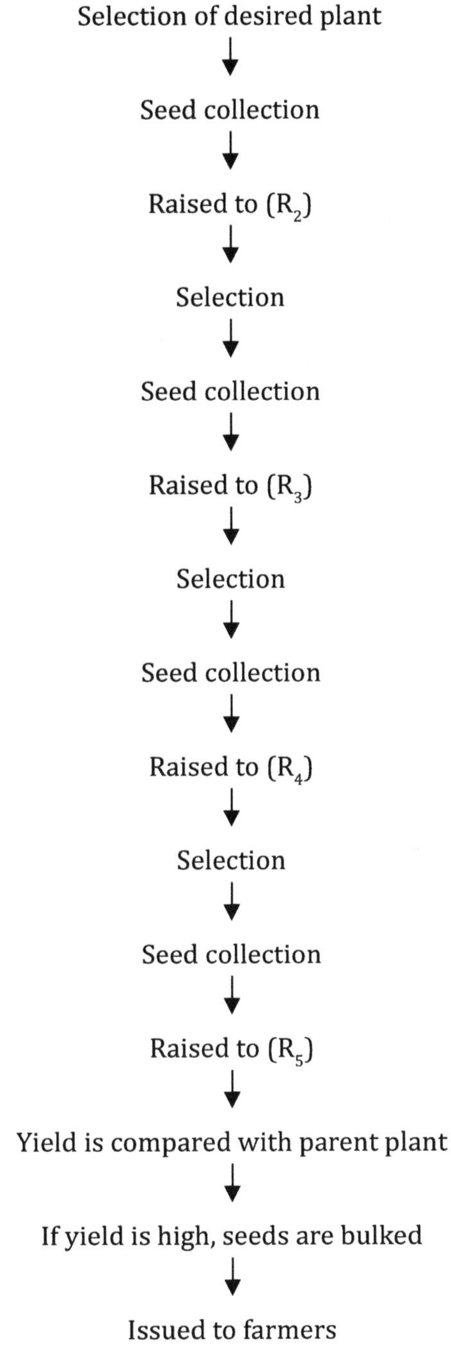
Selection of desired plant
Seed collection
Raised to (R_2)
Selection
Seed collection
Raised to (R_3)
Selection
Seed collection
Raised to (R_4)
Selection
Seed collection
Raised to (R_5)
Yield is compared with parent plant
If yield is high, seeds are bulked
Issued to farmers

f. Rootstock Breeding

The rootstock breeding differs from fruiting cultivar development in the length of breeding cycle and in the history and scope of improvement. Breeding rootstock for fruit crops is slower than scion breeding in the same species. This is due to testing requirements of rootstocks that reduce the opportunity for comprehensive first tests on individual plants and to expanding selection criteria for rootstock. Rootstock breeding is a relatively new discipline of fruit crops improvement and novel function of rootstocks still are being understood and developed.

g. Heterosis Breeding

Heterosis refers to the superiority of F_1 hybrids in one or more characters over both of its parents. The term hybrid vigour is used as synonym for heterosis. It is increased vigours, growth and yield or function of a hybrid over the parents, resulting from crossing of genetically unlike organism. Heterosis differs from luxuriance. The term heterosis was first used or coined by Shull in 1914. In some cases hybrid may be inferior to weaker parent and this is also regarded as heterosis. Some important features of heterosis are given below:

1. Superiority over parents
2. Confined to F_1 hybrids
3. Genetic control
4. Reproducible
5. Association with SCA
6. Effect of heterozygosity
7. Conceals recessive genes
8. Low frequency

Genetic Basis of Heterosis

There are two main theories which have been advanced to explain the mechanism of heterosis. One is the dominance hypothesis and second is the overdominance hypothesis. The epistasis is also considered to be associated with heterosis.

1. Dominance Hypothesis

This theory was proposed by Davenport (1908), Bruce (1910) and Keeble and Pellew (1910). According to this hypothesis, heterosis is the result of the superiority of dominant alleles when recessive alleles are deleterious there the deleterious recessive genes of one parent are hidden by the dominant genes of another parents and the hybrid exhibits heterosis. Both the parents differ for dominant genes.

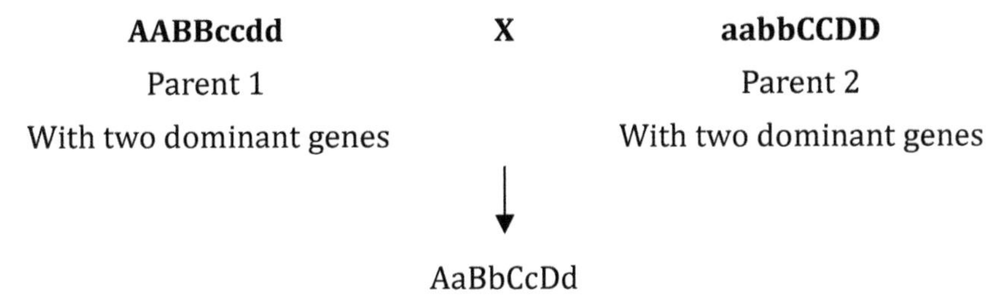

Objection of dominance hypothesis: There are two objections:

- If the hypothesis is true it should be possible to obtain pure heterosis individuals in F_2 which are homozygous for all the dominant genes (Jones in 1917).
- If the heterosis is due to dominance the F_2 curve should be skewed towards dominant genes but the curve F_2 is found always smooth and symmetrical not skewed (Collins in 1921).

2. *Over dominance Hypothesis*

This theory was independently proposed by Shull and East in 1908 and supported by East (1936) and Hull (1945). This theory is called by various names such as stimulation of heterozygosity cumulative action of divergent alleles, single genes heterosis, super dominance and over-dominance. The term over-dominance was coined by Hull in 1945 working on maize. According to this hypothesis, heterosis is the result of superiority of heterozygote over its both homozygous parents.

Difference between Dominance and Over-dominance Hypothesis

Dominance	*Over-dominance*
Heterosis results due to masking effect of dominant desirable alleles over harmful recessive alleles	Heterosis results due to complementation between divergent alleles

3. *Epistasis*

Epistasis refers to interaction between alleles of two or more different loci. It is also known as non-allelic interaction. The non-allelic interaction is of three types, *viz.,* Additive x Additive, Dominance x Dominance and Additive x Dominance. Epistasis particularly that involves dominance effects (Dominance x Dominance) may contribute to heterosis. Epistasis can be detected or estimated by various biometrical models.

Methods for Estimation of Heterosis

It is the mainly done by three methods:

(i) Mid parent heterosis

When the heterosis is estimated over the mid parent, *i.e.*, mean value or average value of the two parents, it is known as mid parent heterosis. It is also known as average or relative heterosis.

Formula

$$MPH = \frac{F_1 - MP}{MP} \times 100$$

where,

F_1 = Mean of F_1

MP = Mean of two parents

(ii) Better parent heterosis

When the heterosis is estimated over the better parent, it is known as better parent heterosis. It is also known as heterobeltiosis.

Formula

$$HB = \frac{F_1 - BP}{BP} \times 100$$

where,

BP = Mean of two parents

The term heterobeltiosis was used by Bitzer *et al.* (1968) to describe the improvement of heterozygote over the better of the cross.

(iii) Standard heterosis

It refers to the superiority of F_1 over the standard commercial check variety. It is also called as economic heterosis or useful heterosis.

Formula

$$SH = \frac{F_1 - Check}{Check} \times 100$$

Heterosis leads to increase in yield, reproductive quality or ability, adaptability, diseases and insect resistance, general vigour, quality, *etc.*

Manifestation or Effects of Heterosis

- Increased yield.
- Increased reproductive ability.
- Increase in size and general vigour.

- ☆ Better quality.
- ☆ Earlier flowering and maturity.
- ☆ Greater resistance to diseases and pests.
- ☆ Greater adaptability.
- ☆ Faster growth rate.
- ☆ Increase in number of plant parts.

Fixation of Heterosis

1. **Asexual reproduction:** It can be easily conserved by vegetative propagation crops, *e.g.*, potato, sweet potato, banana, *etc.*
2. **Apomixis:** In apomixis seed develops without fertilization. Apomictic seed generally develop from maternal diploid cells, *e.g.*, citrus, blackberries, rose, bluegrass and other flowering plants.
3. **Balanced lethal system:** In many species of evening primrose, genetic segregation is almost completely suppressed by a balanced lethal system.
4. **Polyploidy:** It can be fixed by chromosome doubling or polyploidy especially in interspecific and intergeneric hybrids.

Factor Affecting Heterosis

- ☆ Mode of pollination.
- ☆ Genetic diversity of parents.
- ☆ Genetic base of parents.
- ☆ Adaptability of parents.

Heterosis Breeding

Development of hybrid varieties for genetic improvement of yield is referred to as heterosis breeding.

1. **Enough magnitude of heterosis:** There should be enough magnitude of heterosis for its commercial exploitation. It is generally higher in cross pollinated species than in self pollinated species.
2. **High percentage of outcrossing:** There should be high percentage of outcrossing. This is essential for good seed setting.
3. **Floral biology:** The floral biology should permit large scale production of hybrid seed with less expenditure.
4. **Availability of mass selection and self-incompatibility system:** This system helps in reducing the cost of hybrid seed by eliminating the process of hand emasculation.

There should be proper nicking in the flowering time of A and R lines. In case of self-incompatibility the modifier genes which break self-incompatibility should be absent. A line should be is good in stigma receptivity and R line should have high pollen production.

Use of Heterosis in Breeding

- Heterosis is exploited through the development of hybrid.
- It is commercially exploited in seed production of cross-pollinated crops like onion and cucurbits.
- It has been also used in some self-pollinated species such as tomato, brinjal, *etc.*

Hybrid Varieties

Crosses between genetically different plants give rise to cultivars, called hybrid.

- Productivity
- Genetic constitution
- Adaptability
- Application
- Resistance

Merits of Heterosis Breeding

- Hybrid varieties have higher yield potential as compared to synthetics, composites and open pollinated varieties.
- Hybrid varieties are more uniform and attractive than synthetics composites and open pollinated varieties.
- Hybrid varieties can be developed in both self and cross-pollinated species. Whereas, synthetics and composites are relevant tin cross pollinated species only.
- It is possible to reconstitute the hybrid with same genotype. This is not possible in case of composite varieties and open pollinated varieties.

Demerits of Heterosis Breeding

- Fresh seed has to be produced every year because in F_2 the hybrid produces various types due to segregation and recombination. Farmer has to purchase fresh seed every year.
- The seed of hybrid is costlier than synthetics, composites and open-pollinated varieties.
- Cultivation of hybrid requires more input (fertilizers, water, plant protection, *etc.*) to exploit their full potential.
- Production of hybrid requires more technical skill and area as compared to the synthetics composites and open-pollinated varieties.

B. Non-conventional Methods

Biotechnological methods of crop improvement are called non-conventional methods. These are modern methods and hence they are known as non-traditional methods. These methods are based on tissue culture and genetic engineering.

1. **Anther culture:** The culture of immature anthers into embryos and then plants is called anther culture.
2. **Embryo culture:** The culture of immature or mature embryos into plant is known as embryo culture.
3. **Somaclonal variation:** They are called somatic variants. The formation of variant clones from the cultured callus tissue is called somaclonal variation.
4. **Somatic embryogenesis:** These are called artificial seeds.
5. **Somatic hybridization:** Plants raised from hybrid protoplast is called somatic hybrids, *e.g.,* Tangelo and Tangerus.
6. **Molecular assisted selection:** By molecular marker RFLP, RAPD, AFLP, SSRs, *etc.*
7. **Transgenic plants:** Plants containing introduced DNA from unrelated source are known as transgenic plant or genetically engineered plants. These transgenic develop plants are resistant to biotic and abiotic stresses.

Chapter 6

Germplasm of Vegetable Crops

Introduction

The sum total of genes in a crop species is referred to as genetic resources or gene pool or genetic stock or germplasm. In other word gene pool refers to a whole library of different alleles of species. Germplasm or gene pool is the basic material with which a plant breeder has to initiate his breeding programme.

Major features of plant genetic resources or gene pool –

- ✰ Gene pool represents the entire genetic variability or diversity available in a crop species.
- ✰ Germplasm consists of landraces, modern cultivars, obsolete cultivars, breeding stocks, wild forms and wild species of cultivated crops.
- ✰ Germplasm includes both cultivated and wild species and relatives of crop plants.
- ✰ Germplasm collection from the center of diversity, gene bank, gene sanctuaries, farmer's field, markets and seed companies.
- ✰ Germplasm is the basic material for launching a crop improvement programme.
- ✰ Germplasm may be indigenous (collection within country) or exotic (collected from foreign countries).

Kinds of Germplasm

It consists of various plant materials of a crop such as:

1. **Landraces:** Landraces are nothing but primitive cultivars which are selected and cultivated by the farmers for many generation.

Main features of landraces:

- ☆ Landraces were not deliberately bred like modern cultivars. They evolved under subsistence agriculture.
- ☆ Landraces have high level of genetic diversity which provides them high degree of resistance to biotic and abiotic stresses.
- ☆ Landraces have broad genetic base which again provides them wider adaptability and protection from epidemic diseases and insects.
- ☆ Landraces even respond to selection for high yield but to certain extent.

2. **Obsolete cultivars:** Improved varieties of recent past are known as obsolete cultivars. These are the varieties which were popular earlier and now have been replaced by new varieties. These varieties have several desirable characters and constitute an important part of gene pool.
3. **Modern cultivars:** The currently cultivated high yielding varieties are referred to as modern cultivars. Modern cultivars are also known as improved cultivars or advanced cultivars. These varieties have high yield potential and uniformity as compared to obsolete varieties and landraces. Modern cultivars constitute a major part of working collections and are extensively used as parents in the breeding programmes for further genetic improvement in various characters. Hence these cultivars are in great demand. However, modern cultivars have narrow genetic base and low adaptability as compared to landraces.
4. **Advanced breeding lines:** Pre released plants which have been developed by plant breeders for use in modern scientific plant breeding are known as advanced lines, cultures and stock. They include advanced cultures which are not yet ready for release to farmers.
5. **Wild forms of cultivated species:** Wild forms of cultivated species are available in many vegetable crops. Such plants have generally high degree of resistance to biotic and abiotic stresses and are utilized in breeding programme for genetic improvement of resistance to biotic and abiotic stresses. They can easily cross with cultivated species however wild forms of many crop species are extinct. Moreover entire range of diversity of available wild forms is rarely tapped.
6. **Wild relatives:** Those naturally occurring plant species which have common ancestry with crops and can cross with crop species are referred to as wild relatives or wild species. It is important source of resistance to biotic (diseases and insects) and abiotic (drought, cold, frost, salinity, *etc.*) stresses. However wild relatives are used as the last resort in crop improvement programmes because their use in crossing leads to, (i) Hybrid sterility, (ii) Hybrid inviability, (iii) Transfer or several undesirable genes to the cultivated species along with desirable alleles. This group constitutes a minor part of gene pool including inter-specific derivatives.

7. **Mutants:** Mutation breeding is used when the desirable characters are not found in the genetic stock of cultivated species and their wild relatives. Mutations do occurs in nature as well as can be induced through the use of physical and chemical mutagens. Mutants for various characters sometimes may not be released as a variety, but they add to the gene pool.

Types of Seed Collection

Based on the use and duration of conservation, seed collections are of three types:

1. **Base collection:** Base collection include maximum number of accessions (samples) available in a crop. These are meant for long-term conservation (up to 50 years or more) and are stored at -18 or -20°C in hermatically sealed containers. The seed are dried to 5±1 per cent moisture and have more than 85 per cent initial seed viability and these collections are distributed only for the purpose of regeneration. These are used only when germplasm from other sources is not available for use in breeding. It is also known as principal collection and refers to the whole collection.
2. **Active collection:** This category of germplasm is actively utilized in breeding programmes and these are conserved for medium term (8-10 years or more). These collections are stored at 0°C with moisture content around 8 per cent. Germination test is carried out after every 5-10 years to assess the reduction in seed viability.
3. **Working collection:** There collection are frequently utilized by breeders in their crop improvement programmes. These are stored for short term (3-5 years). The seed is stored at 5-10° C with moisture content of 8-10 per cent.
4. **Core collection:** It refers to a subset of base collection which represents the large collection or base collection. In other words core collection is a limited set of accessions derived from existing germplasm collection, chosen to represent the genetic spectrum in the whole collection.

Germplasm Activities

There are six important activities related to plant genetic resources.

1. Exploration and Collection

Exploration refers to collection trip and collection refers to tapping of genetic diversity from various sources and assembling the same at one place.

It is scientific process and into accounts six important items.

(a) **Source of collection:** (i) Centres of diversity (ii) Gene banks (iii) Gene sanctuaries (iv) Seed companies (v) Farmers fields.

(b) Priority of collection: In which fix priority of collection some area of diversity has been threatened more than other by the danger of extinction. Hence endangered areas and species should be given priority for germplasm collection.

(c) Agencies of collection: The task of germplasm collections are undertaken by crop research institute and agriculture university in collaboration with NBPGR (New Delhi) for indigenous collection and for global collection by IPGRI (Italy, Rome).

(d) Method of collection: Four principal methods are used, *viz.,* (i) Through expeditions to the area or regions of genetic diversity. (ii) By personal visit to gene bank centre (iii) Through correspondence (iv) Through exchange of material.

(e) Method of sampling: There are two methods (i) Random sampling- It is effective in capturing of alleles for biotic and abiotic stresses (ii) Non-random or biased Sampling- It is useful in collection of morphologically distinct genotypes.

(f) Sample size: The sample size should be such that 95 per cent of the total genetic diversity can be captured from the area of collection. To achieve this goal, 50-100 individuals should be collected per site with 50 seeds per plant.

Merits

- ☆ Collection helps in tapping crop genetic diversity and assembling the same at one place.
- ☆ It reduces the loss of genetic diversity due to genetic erosion.
- ☆ Sometime, we get material of special interest during exploration trips.
- ☆ Sometime we come across a new plant species during the process of collection
- ☆ Collection also helps in saving certain genotypes from extinction.

Demerits

- ☆ Collection of germplasm especially from other countries sometimes leads to entry of new disease, new insect and new weeds.
- ☆ Collection is a tedious job. The collection has to be made generally from uncultivated area like hills, mountains, river valleys and forests where the collector faces problems of boarding, lodging and transportation.
- ☆ In the remote areas the collection, sometimes has to encounter with wild animals like elephants, rhinos, tigers, lions and snakes which involves risk of life.

2. Conservation

Conservation refers to protection of genetic diversity of crop plants from genetic erosion. There are two important methods.

(a) In-situ Conservation

Conservation of germplasm under natural habitat is referred to as *in-situ* conservation. It requires established of natural or biosphere reserves national park or protection of endangered area or species. In this method of conservation the wild species and the complete natural or semi-natural ecosystems are preserved together.

Advantages

- ☆ It is a cheap and convenient way of conserving biological diversity as we play a supportive role only.
- ☆ In order to ensure the survival of the species we protect the entire natural habitat or the ecosystem.
- ☆ In a natural system organisms not only live and multiply but evolve as well. A natural ecosystem allows free play of natural agencies like drought, storms, snow, *etc.*

Disadvantage

- ☆ Each protected area will cover only every small portion of total diversity of a crop species. Hence several areas will have to be conserved for a single species.
- ☆ The management of such areas also poses several problems.
- ☆ This is a costly method of germplasm conservation.

(b) Ex-situ Conservation

It refers to preservation of germplasm in gene banks. This is the most practical method of germplasm conservation.

Advantages

- ☆ It is possible to preserve entire genetic diversity of a crop species at one place.
- ☆ Handling of germplasm is also easy.
- ☆ This is a cheap method of germplasm conservation.

Disadvantages

- ☆ Captive populations have limited genetic diversity.
- ☆ The organisms are living outside their natural habitate.
- ☆ It can be difficult and expensive to create and sustain the right environment.

The germplasm is conserved either by:

(i) In the Form of Seed

It is very easy method requiring minimum space, seed can conserved under long terms (50-100 years) medium term (10-15 years) and short term (3-5 years) storage condition.

- **Orthodox seed:** Low moisture, low temperature, without losing viability, *e.g.*, carrot, beet, papaya, peeper, beans, brinjal and cole corps are stored at low temperature for long time.
- **Recalcitrant seed:** Seed show very drastic loss in viability with a decrease in moisture content below 12-13 per cent, *e.g.*, cocoa, coconut, mango, tea, coffee, rubber, *etc.*

(ii) In the Form of Meristem Culture

For conservation of meristem culture, meristem or shoot tip banks are established.

Advantages

- Exact genotypes can be conserved indefinitely free from virus or other pathogen and without loss of genetic integrity.
- It is best for vegetatively propagated crops like potato, sweet potato, cassava, *etc.*
- Vegetatively propagated material can be saved from natural disaster or pathogen attack.
- Long regeneration cycle can be envisaged from meristem cultures and it is very easy.
- Perennial plant which takes 10-20 years to produce seeds can be preserved any time by meristem culture.
- Plant species having recalcitrant seeds can be easily conserved by meristem culture.

3. Evaluation

Evaluation refers to screening of germplasm in respect of morphological economic, biochemical, physiological, pathological and entomological attributes.

- To identify gene sources for resistance to biotic and abiotic stresses, earliness, dwarfness, and productivity and quality characters.
- To classify the germplasm into various groups.
- To get a clear picture about the significances of individual germplasm line.

The evaluation of germplasm is done in three different places:

- In the field.
- In the greenhouse.
- In the laboratory.

4. Documentation

It refers to computation, analysis, classification, storage and dissemination of information. In plant genetic resources, documentation means dissemination of information about various activities such as collection, evaluation, conservation, storage and retrieval of data. Now the term documentation is more appropriately known as information system.

- It provides information about various activities of plant genetic resources like latest characterization, conservation, distribution and utilization.
- It helps explores, evaluators and curators in the conservation of genetic resources
- It helps in making genetic resources accessible to plant breeders and other users.

5. Distribution

The distribution of germplasm is one of the important activities of genetic resources centre.

- Distribution of germplasm is the responsibility of the gene bank centres where the germplasm is maintained and conserved.
- The germplasm is usually supplied to the workers who are engaged in the research work of a particular crop species.
- Germplasm samples are generally supplied free of cost to avoid cumbersome work of book keeping.
- The quality of seed samples to be sent is usually small and depends on the availability of seed material and demands received for the same and several other factors.
- Without distribution to actual users, there is no point in collecting the germplasm.

6. Utilization

It refers to use of germplasm in crop improvement programme. Germplasm can be utilized in various ways.

(a) Cultivated Germplasm

It can be utilized in three main ways (i) As a variety (ii) As a parent in the hybridization (iii) As a variant in the gene pool. Some germplasm lines can be released directly as varieties after testing. If the performance of an exotic line is better than a local variety, it can be released for commercial cultivation.

(b) Wild Germplasm

It is used to transfer resistance to biotic and abiotic stresses. It imposes three main problems (i) Hybrid inviability- inability of a hybrid to survive (ii) Hybrid sterility- inability of a hybrid to produce offspring (iii) Linkage of undesirable characters with desirable ones.

Organizations associated with germplasm:

1. IPGRI/IBPGRI- Rome, Italy in 1972.
2. NBPGR- ICAR, New Delhi in 1976.
3. Quarantine- Directorate of Plant Protection Quarantine and Storage- Faridabad (Haryana), India.

Chapter 7

Biotechnology in Vegetable Crops

Introduction

The term biotechnology has been defined in various ways depending on its fundamental meaning and diversified applications. Sometimes, this term has been defined rather, loosely. A few definitions are as follows:

Biotechnology is the application of microorganism and biological systems to the production of goods and services that are beneficial to human welfare. It is an integration of several discipline including microbiology, biochemistry, genetics, and biochemical engineering (Malik, 1989). In other word biotechnology is the application of recombinant DNA, cell and tissue culture, and other methods used to develop new and improved plants and plant products.

Biotechnology consist of a gradient technologies ranging from the long established and widely used techniques of traditional biotechnology (*e.g.*, food fermentation, biological control) through to modern biotechnology, which is based on the use of new techniques of recombinant-DNA technology (often called genetic engineering), monoclonal antibodies, and new cell and tissue culture methods.

Major Uses of Biotechnology

1. Production of plants (vegetables) rapidly on large scale by tissue culture techniques.
2. Production of disease free materials by meristem culture.
3. Somaclonal variation to broaden the gene base for the isolation of resistance at cell level to disease, soil salinity and acidity, *etc.*
4. To add desirable gene to recipient species through vector like plastids, viruses, plasmids, *etc.*

5. Gene closing of desirable materials for the incorporation in common cultivars.
6. The somatic hybridization in case of distant species.
7. Embryo rescue culture for making distant crosses.
8. Industrial uses.

Micro-propagation

Different techniques of *in-vitro* culture must solve two basic problems. Firstly, the maintenance in a sterile or aseptic condition, *i.e.*, free from microbes cultured cells/organs must be ensured by providing the proliferation and proper development of the cultured cells/organs by providing the necessary nutrients, growth regulators and the other growth conditions. While providing the necessary nutrients, growth regulators and general laboratory facilities, a small piece of the plant to be cultured (referred to as the ex-plant) is removed from a healthy, well-maintained stock/mother plant. The source of tissue will vary from species to specics, but shoot-tips, leaves, stems, lateral buds, and flower tissues are the most commonly used ex-plants. The type of ex-plant chosen depends mainly on the objective of the research worker and the species with which the work is being carried out. In many cases, the choice of the ex-plant may be critical to the success of the study, *e.g.*, immature zygotic embryos for the initiation of embryogenic callus/suspension cultures.

The selected ex-plant is rinsed with a dilute detergent (*e.g.*, Tween, Cetrimide, *etc.*) so followed by surface sterilization using a suitable chemical treatment, *e.g.*, with 0.1 per cent mercuric chloride. After this, the ex-plant is rinsed 4-6 times with sterile distilled water, and placed in a vessel containing the selected culture medium; these entire operations arc carried out under aseptic conditions. The ex-plant may undergo shoot proliferation directly by lateral bud break (if pre-existing meristems are present), or the tissue may undergo a phase of unorganized growth, *i.e.*, callus formation, prior to shoot differentiation. The pattern of growth of the cultures is principally determined by the plant growth regulators (mainly the types and the concentrations of auxin and cytokinin) supplemented in the tissue culture medium.

Stages in Micro-propagation

The main objective in micro-propagation is to produce large number of genotypically identical plants in a relatively short period of time. Thus the micro-propagation may be viewed scientific extension of traditional method of cutting: it employs strict aseptic conditions, adequate nutrients and a controlled environment. Initially, Murashige (1974) divided micro-propagation into three stages, but later researchers have divided the whole process into 4 stages described below.

Stage I. Establishment of Aseptic Cultures

This stage consists of selection of the mother plants (source of the ex-plant) and harvesting surface sterilization and inoculation of the ex-plants. Generally, choice of the ex-plants depends on the species to be micro-propagation and the path of micro-propagation to be followed. The season and growth stage of ex-plant may affect the ease of culture establishment. In most cases growing plant parts give good response in culture. In case of tree species, it is desirable to select about one-year-old branches growing from as close to the base of the tree as possible be necessary to cut off some old branches from close to the main trunk to force the growth of such young branches. This stage primarily requires transfer of ex-plants to the culture en-contaminant free states, followed often by growth cultured, and the contaminated cultures are discarded. This stage culture establishment with a fairly large number of ex-plants and harvesting ally, choice of the ex-plant depends on planted, and the path of micro-propagation to be followed. The seasons, actively initiation initially, several batches need to be cultured and the contaminated cultures are discarded.

Stage II. Growth and Differentiation of Shoots

Micro-propagation is generally based on induced multiplication of the axillary meristem so that each meristem that is either presents in the ex-plant or develops in the culture subsequently use a potential plant. In addition to (a) an existing apical/axillary meristem (b) the shoot on denovo regeneration of adventitious meristems (organogenesis) (c) plantlets may be obtained via somatic embryos. In commercial practice, the four most commonly propagations are enhanced axillary branching (i) in shoot-tip (ii) nodal cultures (iii) regeneration of adventitious shoots (iv) somatic embryogenesis. For axillaries shoot proliferation and adventitious shoot regeneration usually suitable concentration of a cytokinin (generally BA in some cases, a combination of a cytokinin and a low concentration of generally induced by a relatively higher concentration of an auxin (2,4-D is preferred). But once embryogenic cultures are obtained they are maintained on a medium containing a relatively low auxin level.

Stage III. Root Induction in the Shoots

In most commercial operations, the preferred method in this stage is non-sterile rooting. Micro-cuttings are harvested (usually under non-sterile conditions) and stuck in a clean (but non sterile) mixture of vermiculite or in a peat moss plug. If required, the cut ends of shoots may be treated with a suitable rooting mix or a suitable concentration of an auxin like IBA or NAA. These plugs are then placed in mist/fog chamber or polyhouse. However, in laboratory, the root induction is achieved in a suitable culture medium and the rooted ex-plants are then transferred into pots filled with soil rite. Rooting stage is very important part of a commercial multiplication system. On a small seal, shoots are usually rooted on suitable rooting medium containing lower salt concentration than the normal tissue culture medium. Some species root easily in the culture, but for most species a suitable rooting treatment (usually, a low concentration of IBA or NAA) is needed. Sometimes, it may be necessary to allow the shoots to elongate prior to rooting.

The Roles of the Various Growth Regulators Commonly Used in Plant Tissue Cultures

Growth Regulator	Role in Tissue Culture System	Common Examples	
		Natural	Synthetic
Auxin	Adventitious root formation	IAA	NAA
	Adventitious shoot formation (in some species)		2,4-D
	Induction of somatic embryos (usually, 2,4-D)		2,4,5-T
	Initiation and proliferation of callus and suspension cultures		IBA
Cytokinin	Adventitious shoot formation		
	Axillary shoot bud proliferation by overcoming apical dominance	Ziatin	Kinetin
	Cell division in explants and cultures (in conjunction with an auxin)	BA	TDZ
Gibberellins	Shoot elongation	GA_1, GA_7	
	Breaking seed dormancy		
Abscisic acid	Promotion of shoot bud and somatic embryo regeneration	ABA	
	Maturation of somatic embryos		
	Development of dormancy		
Ethylene	Ripening of fruits	Ethylene	
	Senescence of flowers and leaves		
Polyamines	Promotion of shoot formation	Spermidine	
	Promotion of somatic embryogenesis	Spermine	

Stage IV. Preparation for Growth in Natural Environment (Hardening)

The shoots/micro-cuttings/plants, derived in stage III are very small and are not capable of self-supporting growth in soil. So in this stage the efforts are made to condition the plantlets, so that they can grow in external environment and carry out the photosynthesis. After this the transfer of plantlets soil is also very important step, as any mistake in this critical phase can lead to significant loss of propagules. Shoots developed in culture are grown under high humidity and low light intensity leading to less cellular wax. This is the reason the tissue-cultured plants lose water rapidly in external environments. In practice, the rooted plants are removed from vessels and their agar is washed thoroughly, now the small plants are transferred to sterile rooting medium (Vermiculite/Peat +Sand) and are kept in high humidity and less light for several days. Then plants are gradually shifted to light, hardened and transferred to soil. The whole process is known as hardening or acclimatization.

Applications of Micro-propagation

- ✰ Relatively high multiplication rates can be achieved as compared to the conventional vegetative propagation procedures. For example, annual rates of multiplication in bulbs are generally below 5, whereas, micro-propagation by adventitious shoot formation can give multiplication rates of 10 to 1000 per annum. In some cases, rapid clonal propagation rates of 100,000 to 3,000,000 have been achieved through multiplication from apicalor axillary shoot.

- Propagation of such species that cannot be propagated by conventional vegetative propagation methods can become possible. For example, oil palm (Elaeis guineensis has been propagated from callus initiated from apical shoot tissue.
- Parents of outstanding F_1 hybrids can be multiplied and used for hybrid seed production; this is particularly useful in such as where the parents are difficult to multiply because of self-incompatibility (*e.g.*, cauliflower).
- The technique permits propagation of high-value unique lines, *e.g.*, seedless, gynoecious species, male sterile lines, *etc.*
- Tissue cultures are also useful for maintaining special genotypes used in breeding, this attribute is particularly important if the performance of progeny from a cross is not known until several years later (*e.g.*, forage species like alfalfa or tree species).
- In addition, *in-vitro* propagation alone or together with heat therapy can be used to produce disease free plant materials, which is preferred for international germplasm exchange.
- Round the year propagation is possible without environmental effects.

Limitations of *In-vitro* Propagation

The cost of production per unit of propagule can be non-competitive with conventionally propagated material because of collaborate facilities required, and the maintenance and labour cost. There are many species for which, shoot multiplication or regeneration is still difficult or even impossible. At the same time, some variation may be observed in the plants obtained through *in-vitro* techniques as compared to the original mother plant. These variations may arise from (a) epigenetic charges, which are short-lived (usually only one generation) developmental abnormalities carried from the *in-vitro* culture conditions over to the field, *e.g.*, strangely-shaped leaves or whole plants. But many of the variations may result from (b) stable genetic changes, *i.e.*, soma clonal variations, arising during tissue culture: the frequency of such variations increases with the culture duration and is affected by several tissue culture parameters. These features are very important for ornamental plants as they are sold directly after Stage IV, *i.e.*, hardening.

In case of perennial species, mature trees are preferred for cloning because they have been in the field long enough to prove their value. But mature trees are often rather difficult to propagate in some cases, it may be possible to overcome this problem by subjecting the selected trees to appropriate horticultural manipulations that rejuvenate the plants or provide such explants that are quite responsive to tissue culture. The available literature on micro-propagation indicates that much attention is paid to the nutritional requirements of the cultures, while culture conditions are invariably the most neglected factor.

Tissue Culture

Tissue culture is a term used to indicate the asceptic culture *in-vitro* of a wide range of excised plant parts. It is used for both propagation and modification of genotypes (production of somaclonal variation and double haploids), biomass production of biochemical products, plant pathology, preservation and storage of germplasm, *etc.* micro-propagation and tissue culture are also used interchangeably. Tissue culture begins with the excision of a small piece of plant, freeing it from micro-organisms and placing it into asceptic culture. The term used for small piece of plant initially excised is ex-plant.

History of Plant Tissue Culture

The emergence of theory of totipotency laid to the development of different *in-vitro* techniques. Totipotency may be defined as the ability of a single differentiated mature cell to develop into a complete plant, preferably via formation of an embryo. Haberlandt was the first to culture, in 1902, fully differentiated cells isolated from leaves of several plant species. He observed that t cells survived and underwent enlargement, but they failed to divide. However, the first successful report of a continuously growing *in-vitro* culture of a plant organ (tomato root tip) was that by P.R. White in 1934. At the same time, Gautheret was able to continuously grow the cultures from carrot root cambium. Thus, Gautheret and White independently reported the establishment of growing cultures of plant tissues/organs and they laid the foundation for further research in also field.

In 1904, Hanning initiated a new line of research, which has now become the technique of choice for the rescue of inter-specific hybrids. He cultured nearly mature embryos of some cruciferous plants and successfully grew them to maturity. This relatively simple demonstration paved the way for now well-established and widely used technique of embryo culture. In 1925, Laibach demonstrated a potent practical application of embryo culture in the field of plant breeding. He isolated almost fully developed embryos from the non-viable virtually endosperm less seeds of an inter-specific cross (*Linum pereme* x *Linum austriacum*) and cultured on a simple nutrient medium and obtained viable seedlings.

Tissue Culture Techniques

In such techniques plant cell/tissues/organs are cultured to get entire plant. Different plant parts used are meristem tips (onion), young tips (garlic), floral meristem (cauliflower), young expanded leaves and cotyledonary tissue (lettuce), excised hypocotyle segment (tomato), root segment (brinjal), shoot tips (sweet potato and cucumber), anthers (capsicum, tomato, potato and Brussels sprout), *etc.*

Tissue Culture Techniques in Vegetables

Plant Organ Used	*Vegetables*
Callus/tissue culture	Lettuce, sweet poatato and coriander
Cell culture	Garlic
Meristem culture	Onion and cauliflower
Organ culture	Lettuce, tomato, brinjal, sweet potato and cucumber
Embryo culture	Tomato, okra, french bean, brassicas and pumpkin
Anther and pollen culture	Pototo, tomato, pea, asparagus, brassicas and capsicum

The Laboratory Setup

Cleanliness is the major consideration when planning a tissue culture laboratory as the losses from contamination are reported to be up to 50 per cent. Thus when we consider high value product (commercial production), loss from contamination are unacceptable. The commercial laboratories should have easy-to-wash floors and walls, such as those made from acrylic or urethane. The traffic pattern and workflow must be considered in order to maximize cleanliness. The cleanest rooms/areas are the culture room/primary growth room and asceptic transfer area; thus it is the best to keep these areas separate so that they are not entered directly from outside. Ideally, the media preparation area should lead to sterilization area then to aseptic transfer area which eventually should be connected with the culture room. There are a variety of equipments available for tissue culture laboratory, which are usually located in the preparation room.

Media Preparation Area

A wide variety of equipments are used in tissue culture experiments; these equipments are generally located in the media preparation area. This room has the facility for media preparation and sterilization (auto-claving). It also has a separate facility for cleaning of glassware and storage of the chemicals. The basic equipments in this area are a precision balance (for weighing of chemicals), a pH meter (to measure and adjust media pH), a refrigerator (for storage of chemicals/stock solutions), a water purification system (for obtaining distilled water), a magnetic stirrer (for mixing of solutions) and a hot plate (for heating of media/solutions).

The aseptic transfer room is an area where any operation requiring sterile condition is performed. The aseptic transfer area needs to be as clean as possible to avoid chances of contamination. It is preferable to have a separate room for aseptic transfer; this decreases spore circulation. This area has the laminar air-flow workstation to be used for all aseptic works. Ultra-violet (UV) lights are sometimes installed in the transfer areas to disinfect the room; these lights should only be used when people and plant material are not in the room. Surfaces inside the aseptic transfer area should be smooth to minimize the amount of dust on them. The aseptic transfer area also has a burner, forceps, seal, balance, microscopes and often a centrifuge.

Growth Room

The growth room or culture room is the area where cultures are kept, following inoculation to enable their maintenance and to support the desired growth and development. Generally, this area is separated from rest of the laboratory and is well-cleaned. The most important facility in the culture room is the culture racks fitted with tube lights and a photoperiod regulator. Temperature is another important concern in the culture room so that the culture room has the facility for temperature control. Usually, a temperature of 25±2°C is maintained through air conditioners and heaters. Continuous power supply is a must since a failure of electricity will ruin the experiments looking in to the value of products from a commercial set-up a humidifier/dehumidifier is also placed in the culture room in order to maintain the relative humidity.

Acclimatization Area

The plants regenerated under *in-vitro* conditions are not hardy enough to withstand the vigours of the environment if they are directly planted in the open field. Thus an area is needed where plantlets are first conditioned to make them ready for field transfer; this area is called acclimatization area. The acclimatization area is a high humidity area with sprinkler irrigation facility. It may be in the form of greenhouse/polyhouse facility.

Sterilization

Tissue culture media are usually rich in nutrients; therefore, they support prolific growth of many micro-organisms (*e.g.*, bacteria, fungi, *etc.*). On reaching the medium, these micro-organisms generally, grow much faster than the cultured plant tissues so that they over when them. In view of this, suitable sterilization procedures should be employed to minimize, as much as possible. The risk of microbial contamination of the cultures to achieve this, all the materials, *e.g.*, culture vessels, instruments, medium, plant materials, *etc.*, as well as the working area should be made free of microbes. This can be achieved by the following approaches of sterilization, *e.g.*, dry heat, flame sterilization, autoclaving, filter sterilization, surface sterilization, *etc.*, each of these procedures is suited for some but not other items of the culture system.

Different Sterilization Procedures and Material to be Sterilized by each Procedure

Sterilization Procedure	*Material to be Sterilized*
Dry heat (160-180°C for 2-3 hr.)	Glassware (culture tubes. petri plates, pipettes, *etc.*)
Flame sterilization	Scalpels, needles, forceps, culture vessel mouths
Autoclaving (121°C for 15-20 minutes)	Culture media, glassware, contaminated cultures
Filter sterilization (0.22 mm membrane filter)	Heat labile compounds (GA_3, ABA, Zeatin)
HEPA filter	Air (blown through laminar air flow workstation)
70 per cent ethanol wipe	Surface of laminar air-now bench, operator's hands, *etc.*

Nutrient Medium

All the nutritional requirements for the normal growth and of plant cells and tissues are met by the nutrient or culture medium, which is often simply referred to as medium. A culture medium generally contains organic and inorganic nutrients, certain vitamins, required grow regulators as per the experiment and sometimes certain amino acids. Mineral elements play a vital role in the development of a plant as many of the cements form a part of important bio molecules For example, magnesium, forms a part of chlorophyll and nitrogen is a component and of amino acids and nucleic acids. The media contain six macro-nutrients (N, P, K, Ca, Mg and S) and six micro-nutrients (Fe, Mn, Cu. Zn, B and Mo); the former are needed in concentration of greater than 0.5 mmol, while the latter are required at levels lower than 0.5 mmol. All these elements added in the form of their salts (*e.g.*, Ca and Mg as their sulphate salts, *etc.*). The iron is added in the form of $FeSO_4$- EDTA; this increases the availability and stability of the complex, particularly at the lower pH values. In addition, vitamins like nicotinic acid pyridoxine HCI, thiamine HCl, and mistily; folic acid may also be added in some cases. Since the concentration of media constituents is very low, concentrated stock solutions micro-elements (except Fe), Fe-EDTA, vitamins and separately for each growth regulator. For preparation of the media, double distilled water and analytical grade chemicals are prepared should be used any stock solution showing bacterial/fungal contamination or any precipitate is discarded.

In most of the experiments, sucrose is used as the carbon and maltose is used rarely. In specific cases, complexes of undefined composition (*e.g.*, casein hydrolysate, coconut milk, yeast extract, and tomato juice, *etc.*) may be added in the culture medium. Solidified media provides a substratum to the growing static tissue. For gelling solidification of the medium, agar (a polymer of sulphate galactose obtained from red algae Gelidium) is added to the medium at 8-10 g. The unique property of agar is that at temperature above 60°C it becomes liquid and it solidifies below 45°C to form a semi-solid gel. Gel rite and phytagel serve as a good substitute of the agar; they are used at a lower concentration (0.2 per cent). If there is a problem of browning of medium due to the phenols secreted by the cultured cells tissues, activated charcoal is added to the medium. As the stability and absorption of the ions from the media is affected by pH, each medium is set to a particular pH; most of the media are set to a pH around 5.5.

Growth regulators, which are used to supplement the medium, are divided into six major groups, *viz.*, auxins, cytokinins, gibberellins, abscisic acid, ethylene and polyamines. In the auxin group, the main growth regulators are, naphthalene acetic acid (NAA), Indole 3-butyric acid (IBA). Indole 3-acetic acid (IAA), 2, 4-dichlorophynoxy acetic acid (2,4-D), and P- chlorophenoxy acetic acid (PCPA). The cytokinin group is dominated by 6-benzyl adenine (BA) and Kinetin and thidiazuron (TDZ), Zeatin and 2-ip (isopentenyl adenine) are also used occasionally. GA; is perhaps the sole gibberellins to be used in tissue cultures and abscisic acid is used to achieve some specific objectives.

Meristem Culture

This term describes the culture of ex-plants bearing intact apical/axillary shoot meristem with a view to obtain complete plantlets. The apical meristem (*i.e.*, growing shoot-tip) could be cultured *in-vitro* resulting in continued and organized growth; this is sometimes known as meristem tip culture, which gives enhanced axillary branching that minimizes the risk of somaclonal variations. These shoot apices finally give rise to small shoots, which can be rooted. Meristem tip culture is of practical significance and can be utilized to produce virus/disease free plants, shoot meristems are generally free from viruses; to achieve this, the apical meristem that is about 100-200 μm needs to be cultured. But for micro-propagation, longer shoot tips of about 2 to 5 mm can be cultured. The technique involves manipulation of culture conditions so that meristem could develop in the organized state to form shoots and later roots to produce complete plantlets. The first step in meristem culture is to establish contamination free cultures of the meristem ex-plants. The multiplication rate is also slow at first, but later if the culture conditions are suitably optimized much higher multiplication rates could be achieved if the growth regulators are appropriate.

Cell Culture

Among the different technique available, cell culture provides an opportunity to study the behaviour and potentialities of plant cells under *in-vitro* culture condition. Since the *in-vitro* raised callus is grown under aseptic condition, it can be good source material for isolation of single cells, but plant part such a leaves are also a good starting material single cells can be isolated either by mechanical method or by using an enzyme. The isolated cells can be culture as a single cell or as a suspension culture. The single cells can be cultured by the micro-chamber technique or by Begmann single cell planting technique.

Somatic Embryogenesis

In 1985, Steward and Renert were the first to discover somatic embryogenesis in *Daucus carota* cell culture. Van Arnold has defined that it is a process in which a bipolar structure resembling a zygotic embryo develops from non-zygotic cell without vascular connections with the original tissue. Thus it is a multi-step process starting with the formation of pre-embryonic mass to the ex-plant cultured on media supplemented with the required growth regulators (general auxin like 2,4-D and/ or, in some cases, a cytokinin). This is followed by initiation and proliferation of embryos on solidified or liquid media. In some plants, a medium with high sucrose concentration is required to promote normal somatic embryogenesis development, *e.g.*, formation of scutellum (storage organ) in maize, likewise, somatic embryos of red oak germinated only after treatment with osmotically active sugars. Finally, the somatic embryos are cultured for their maturation on a medium with increased osmotic potential and ultimately the somatic embryogenesis germinates to produce plantlets.

The main applications of somatic embryogenesis include, provision of cell lines for genetic engineering, long-term storage of a particular genotype and in understanding of genetic and developmental phenomenon of the cells. Somatic embryogenesis has been used for micro-propagation of some plant species like oil palm on a commercial scale.

Somaclonal Variation

Exploitation of the natural variation that exists in plant populations has led to the development of the varieties and hybrids that we have today. When regenerates from tissue culture closely observed, genetic off-type were often noticed among these plant; these were called somaclonal variants and the phenomena was referred to as somaclonal variation. This was seen as a novel source of variation that could be exploited for improvement. However, somaclonal variation would be limiting in the commercial scale.

The frequency of off-types occurring in culture varies with species, culture type and the age attributed to a number of chromosomal and genetic alterations that occur in the cultured cells. However, the cause of somaclonal variation is not fully understood, but somaclonal aberrations, errors by DNA polymerase and transposition, *etc.*, may be involved. Thus far it is not possible to control somaclonal variation, but is could be minimized by reducing the duration of *in-vitro* culture and by avoiding a callus phase. Further, a proportion of the somaclonal variations can be epigenetic and unstable, which calls in question their value of applied agriculture. Somaclonal variation has been promoted by the use of radiation climate mutagens and by colchicines, a chemical generally used to induce auto-polyploidy, *i.e.*, as a polyploidizing agent. Somaclonal variation offers an opportunity to change one or two characters without altering the remaining genotype of an otherwise desirable variety. This is often not possible using conventional breeding strategies. Somaclonal variation is a big problem in micro-propagation, but commercial companies utilized somaclonal variations for creating usable genetic variability. For example carrot and pepper varieties with better eating quality have been selected from among plants regenerated from callus culture, but in these cases commercial ventures had used these techniques with extreme caution. Initial reports of potentially useful variation included improved tuber shape colour and late blight resistance in potato and increased solids and tolerance to *Fusarium oxysporum* culture filtrate in tomato.

In-vitro selection involves applying a suitable selection pressure to the cell cultures to the cell culture to isolate useful somaclonal variants. For example, resistance to pathogens has been selected for using specific toxins produced by the pathogens as selection agents; the toxins were added to culture medium, and toxin resistant clones were isolated and plants regenerated. Similar approach has been used to select for resistance to certain herbicides, low or high temperature, aluminum and the strategy obviously requires a reliable method of regeneration from callus/suspension cultures. In addition, the tolerance to the selection agent at the cellular level must correspond closely to that in the regenerated plant in the field and transferable to other plants through conventional plant breeding.

Anther Culture and Doubled Haploids

Among the several biotechnological tools available to researchers, culture of anther and pollen, holds special promise because of its potential applications in crop breeding. Natural occurrence of haploids is rather rare and is confined to few species. In 1964, Guha and Maheshwary reported the first success of haploid plantlet: production through embryogenesis from anthers of datura cultured *in-vitro*; this phenomenon is often referred to as androgenesis and the early progress with the techniques of androgenesis was mainly with solanaceous plant species.

The production of plants from haploid pollen grains has been one of the major contributions of plant tissue culture research to plant breeding as it allows rapid recovery of homozygous lines from heterozygous source materials. One application of this method involves production of F_1 hybrids between selected parents having desired traits: the anthers/microspores from the F_1, plants are cultured on a suitable medium to obtain haploid plantlets and the chromosome number of these plantlets is doubled (generally, using colchicines) to recover homozygous doubled haploid plantlets. These homozygous plantlets readily express the recessive deleterious mutations (often difficult to select again: in the segregating generations of crosses) as well as desirable quality traits, which make the selection relatively easy and considerably more efficient.

For anther culture, it is desirable to grow the source plants under controlled conditions initially, flower buds of different sizes are collected and used to determine the developmental stage of the pollen grains present in their anthers; this allows one to correlate the pollen developmental stage with the stage of the flower bud. This facilitates isolation of anthers in the desired developmental stage since androgenesis response depends mainly on the developmental stage of microspores. It has been observed that uninucleate microspores midway between the tetrad release and the first pollen mitosis are the most responsive. The collected flower buds arc surface sterilized under aseptic conditions and opened to remove the anthers that are placed horizontally onto the culture medium. Care should be taken to avoid injury to the anthers since injured anthers are likely to form callus from somatic tissues. If isolated microspores have to be cultured, the anthers are macerated in liquid medium and the debris is removed using a suitable sieve.

Usually, callusing starts within 5-6 weeks of culture in dark and the anthers burst open revealing the callus mass. Once the callus formation is initiated, the cultures are subjected to alternating periods of light and darkness. The callus would regenerate either pollen embryos or shoots that would, ultimately, give rise to new plantlets, which are transplanted into soil. As the plantlets derived from anther culture are expected to be haploid (have only one set of chromosome) they will normally be weak and will not produce viable gamete. Therefore, their chromosome number is doubled by a suitable treatment, *e.g.*, treatment with a freshly prepared solution of colchicines. Generally, the young shoots are treated with 0.2-0.4 per cent colchicines solution and then transferred to medium for rooting the plantlets thus obtained are evaluated for their morphological and reproductive features to determine if their chromosome number has been doubled. The plants attained after chromosome doubling of haploid plants are known as doubled haploid plants and such plants are, as a rule, homozygous for all their genes.

Applications

The first step in hybrid development is to develop inbred parental lines by repeated self-pollination. This can be very slow process in some crop, *e.g.,* in some cruciferous crops. However, by producing doubled haploids one can dramatically produce homozygous lines rather rapidly. The first step in hybrid development is to develop inbred parental line by repeated self, such as onions and carrots, which normally require 2 years to flower and in those crops where self-pollination presents problems. The development of improved pure-line varieties by anther/ microspore culture is already a reality in China. However, most of the Chinese work is related to cereal crops like rice and wheat; some anther culture-derived plants were found to have heavier seeds and higher protein content. Difficulties associated with anther culture are low frequency of haploid production, difficulty in distinguishing spontaneously doubled haploids from diploids regenerating from somatic tissues and the considerable time and labour needed to generate large numbers of plants required for critical evaluation in a breeding programme. Further, a dependable technique for haploid production is not available in the case of many crops, *e.g.,* many of the grain legumes.

Production of Haploid (n) Plant through Anther/Microspore Culture in Horticultural Crops

Crop Species	*Mode of Haploid Development*
Asparagus officinalis	Direct/indirect androgenesis
Beta vulgaris	Direct/indirect androgenesis
Brassica oleracea	Direct/indirect androgenesis
Cupsicum annuun (Fresh World Farms Sweet Mini pepper developed by DNAP Holding Corporation)	Anther culture-derived variety released for commercial cultivation
Cucumis sativus	Indirect androgenesis
Solanum lycopersicum	Direct/indirect androgenesis
Solanum tuberosum	Direct/indirect androgenesis

In case of vegetables, some work has been, done on haploid production. The recent release for commercial cultivation of sweet pepper variety Fresh World Farms Sweet Mini-pepper developed through anther culture confirms the usefulness of this strategy. Fruits of this variety have increased sugar content and deep red colour and are nearly seedless; these are desirable features from them from consumer's viewpoint.

Embryo Rescue

The culturing of embryos *in-vitro* was the first tissue culture technique to be applied directly to plant improvement. In this technique, young inter specific hybrid embryos are excised from developing seeds and are cultured on a nutrient medium to obtain new plantlets; this is called embryo rescue since these hybrid embryos would otherwise have not survived. Hanning, in 1904 cultured mature embryos of Raphanus and Cochlearia using aseptic techniques. Generally, it is difficult to culture embryos before a certain stage of its development, but 2-4 celled embryos of *Brassica juncea* have been successfully cultured. *In-vitro* culture of ovaries, ovary slices, ovules and even isolated embryos of inter specific hybrids often successfully bypass post-zygotic incompatibilities that result from endosperm abortion.

Since the embryos develop in sterile condition (*i.e.*, inside the immature fruit ovary), surface sterilization of the embryo per se is not required. However, It is necessary to surface sterilize the immature fruit before dissecting out the developing seed and the young embryo. If the embryos are large enough to be seen with naked eyes, they are excised with ease. However, a stereoscopic microscope would be needed for dissecting out the smaller young embryos. While culturing the young embryos, a filter paper may be kept between the embryo and culture medium. Another option is to use *in-vitro* cultured endosperm as nurse tissue.

The nutritional requirements of immature embryos are different from those of normal mature embryos as the latter are capable of synthesizing all the bio-chemicals they need from a relatively simple nutrient medium. Some complex additives of undefined nature, *e.g.*, coconut milk and casein hydrolyse and tissue extracts, may sometimes be added into the medium. The growth regulator requirement varies with the stage of embryo development: older embryos may not require growth regulators, but young embryos may need a combination of low concentrations of an auxin and a cytokinin.

Applications

Wide crosses are valuable tools especially for transferring genes governing specific traits like resistance to biotic and abiotic stresses, improved quality, even higher yields, *etc.*, to crops from their wild relatives. However, wide crosses often fail to produce mature viable seeds; in such cases the immature embryos are removed from the developing seeds and cultured in the laboratory to obtain hybrid plants. Thus embryo culture enables crosses of crop species with a greater number of related wild species and, thereby, permits the breeders to access a much wider range of genes for genetic improvement of the crop plants. This method has been successfully used in crops like tomato, brinjal and onion to recovers inter specific hybrids.

Embryo culture can be employed if there is some problem in seed germination due to prolonged dormancy. For example, orchids are difficult to propagate through seed as they do not have any stored food, and in some cases the embryo development is incomplete at the time of seed maturity. In this situation, embryos can be removed from mature seed and cultured onto a suitable nutrient medium to obtain plantlets. Similarly Iris seeds take several years to germinate; this problem can be overcome with embryo culture technique.

Selected Examples of Embryo Rescue in Vegetable Crops

Inter-specific Hybrid	*Hybrid Plants Recovered by*
Solanum lycopersicum x *S. peruvianum*	Embryo rescue
Phaseolus vulgaris x *P. angustissimus*	
Vigna pubescens x *V. unguiculata*	
Solanam melongena x *S. rorvum*	

Protoplast Fusion and Somatic Hybrids

Protoplast fusion is another method for producing genetic variations in the laboratory. A protoplast is a plant cell from which the outer cell wall has been removed; this is generally achieved by treating plant cells with a combination of pectinase and cellulase enzymes. The process of protoplast fusion of two different plant species/varieties and the development of hybrid plantlet from them is known as somatic hybridization and such plants are called somatic hybrids. Another related term is cybrid (cytoplasmic hybrid), it means a hybrid produced by fusion of one protoplast with a cytoplast, which is a protoplast that lacks nucleus. Cybrids provide unique opportunity for transferring plasmagenes from one species to another species in a single generation. Somatic hybrids may contain the complete chromosome complement of the two fusion parents (symmetric hybrids) or may contain full chromosome complement of one fusion parent but one (even a fragment of one) to few chromosomes from the other fusion parent (asymmetric hybrids). Symmetric hybrids between two distinct species are, in fact, themselves novel species and sometimes give rise to useful/promising materials for crop improvement. But asymmetric hybrids provide an opportunity for gene transfer from one species into another.

Somatic hybridization or protoplast fusion involves a series of operations, such as isolation of protoplasts from cells, culturing them on a suitable medium, inducing them to divide and regenerating plantlets from the resulting cell masses. Once the method for regeneration from protoplasts of both (in any case, at least one of) the fusion partners is developed. The fusion experiment is attempted and the fusion products are cultured to regenerate somatic hybrid plants. For isolation of protoplasts, leaf mesophyll cells are an important source material. However, other parts of the plant can also be used for protoplast isolation provided the cells have not undergone lignifications. There are two methods for protoplast isolation, (i) mechanical method (ii) enzymatic.

Applications

Protoplasts derived from genetically different plants can be fused allowing their nuclear and other cellular contents to be combined. The somatic hybrids, therefore, will have traits from both parents. Some novel varieties of Chinese cabbage and other *Brassica* developed through protoplast fusion have been marketed. Protoplast fusion has been used to transfer male sterility (lack of pollen viability) in *Brassica* species (cabbage, cauliflower, broccoli), where it is used in the production of F_1 hybrid seeds. Asymmetric hybrids provide an opportunity for the introgression of desirable genes from one fusion partner to other. The genetic base in some crops may be narrow; in such cases, the genetic base can be widened by synthesizing the species in question from known diploid species. For example, *B. napus* can be synthesized by producing symmetric hybrids between *B. oleracea* var. *botrytis* and *B. compestris* var. *oleifera*. The symmetric hybrids can also be used in a hybridization programme to transfer desirable traits from a related species conventionally, cytoplasm transfer from one strain to another is achieved by the back cross method that requires 6-7 years. However, using the somatic hybridization technique, cyloplasm can be transferred in one year (*i.e.*, by making cybrids)

Protoplast Fusion in Vegetable Crops

Crop	*Method of Protoplast Fusion*
Symmetric hybrids	
Allium ameloprasum + A. cepa	PEG-mediated chemical fusion
Solanum lycopersicum + S. peruvianunm	PEG-mediated chemical fusion
Solanum lycopersicum + S. pennelli	Chemical method using PEG
Solanum lycopersicum x *S. pennelli* hybrid + *S. melongena*	Chemical method using PEG
Solanum lycopersicum x *S. richi*	Chemical and electrical methods
S. melongena cv. *Dourga, S. khasianun*	Electro fusion
S. melongena, S. nigrunm	Chemical and electrical fusion
S. melongena, S. symbriifolium	PEG-induced fusion
S. melongena + S. tervum	Chemical and electrical fusion
Asymmetric hybrids	
Daucus carota + Aegopodium podagraria	
D. carota ± Oryza sativa	
D. carota ± Petroselinum hortense	

There are some limitations of the protoplast culture and fusion technique, *e.g.*, for many of the crops like cereals and pulses the protocol for obtaining somatic hybrids is not available. Sometimes somatic hybrids may show genetic instability and somatic hybrid cells may either fail to regenerate plantlets or give rise to sterile regenerates.

Chapter 8

Breeding for Disease Resistance

Introduction

Stress refers to adverse conditions for crop growth and production imposed by either environmental factors or biological factors. Thus stress is of two types, *viz.,* (i) biotic (ii) abiotic. The stress that is caused by biological agents or factors, such as diseases, insects and parasitic weeds, is known as biotic stress. When the stress is caused by environmental factors or non-biological factors, it is referred to as abiotic stress. Abiotic stress is generally caused by factors like deficiency or excess of nutrition, moisture, temperature and light; presence of harmful gases or toxicants; and abnormal soil conditions such as salinity, alkalinity and acidity. All crop plants suffer from both biotic and abiotic stresses to varying degrees. This chapter deals with breeding for resistance to diseases.

Genetic Resistance

Genetic resistance refers to those heritable features of a host plant that suppress or retard development of a pathogen (a disease causing organism) or insect. In other words genetic resistance is the ability of some genotypes to give higher yields of good quality than other varieties at the same initial level of pest (disease or insect) infestation under similar environmental conditions. Thus resistance is defined in relation to susceptible varieties. Genetic resistance is considered as a major form of biological control of biotic stresses. Main features of genetic resistance are given below:

- Genetic resistance is governed by nuclear genes or cytoplasmic genes or both. In other words, genetic resistance is an inhibit mechanism or inherent property.

- ☆ Genetic resistance is measured in relation to susceptible varieties or genotypes.
- ☆ Breeding of resistant cultivars takes into account the genetic variability of both pests and host plant.
- ☆ The resistant variety may become susceptible after few years due to formation of new races of pathogen or new biotypes of an insect.
- ☆ Breeding for disease and insect resistance differs from breeding for higher yield.
- ☆ Genetic resistance is an effective means of controlling biotic stresses in crop plant and also for minimizing the losses due to abiotic stresses.

Types of Genetic Resistance

There are two types of genetic resistance depending upon the number of races controlled. These are: (1) vertical resistance (2) horizontal resistance. A brief description of each type is presented below:

1. Vertical or Specific Resistance

Specific resistance of a host to the particular race of a pathogen is known as vertical resistance. This type of resistance is governed by one or few genes and therefore, is referred to as oligogenic resistance. When the resistance is controlled by single gene, it is called monogenic resistance. Since vertical resistance controls only one race of a pathogen, it is also termed as specific resistance. Because of its simple inheritance, it is known as qualitative resistance. As the controlling genes have distinct effect, it is also known as major gene resistance. The host with vertical resistance controls only one race; therefore, it is also known as non uniform resistance. Main features of vertical resistance are given below:

- ☆ Vertical resistance displays discontinuous variation among genotypes and classification of genotypes into resistant and susceptible classes is possible.
- ☆ Transfer of oligogenic resistance from one host genotype to another is simple.
- ☆ Oligogenic resistance is usually short lived or less durable. The resistance can easily break down when new race of a pathogen is formed.
- ☆ Vertical resistance provides protection only from one race of a pathogen.
- ☆ It has high heritability and can be easily identified in the breeding programmes.
- ☆ Vertical resistance applies to host pathogen gene for gene relationships.

2. Horizontal or General Resistance

The resistance of a host to all the races of a pathogen is called horizontal resistance. This type of resistance is called by various names as per reasons given below:

- **General resistance:** The host plant provides protection from all the prevailing races of a pathogen.
- **Polygenic resistance:** The resistance is controlled by a number of genes, also called quantitative resistance.
- **Minor gene resistance:** Each gene involved in the resistance has small effect which is not visible.
- **Non-specific resistance:** The host resistance is not for a specific race of a pathogen. The resistance is similar to all the races of a pathogen, hence also called as uniform resistance.

Main features of horizontal resistance are given below:

- Horizontal resistance exhibits continuous variation among genotypes and, therefore classification of genotypes into different distinct classes is not possible.
- It has low heritability and therefore, identification of resistant types is difficult.
- It provides protection from several races of a pest.
- The resistance cannot be easily overcome by new races of a pathogen due to polygenic control.
- It is difficult to transfer polygene resistance from one host genotype to another because individual allele in parent can be identified.
- General resistance is not applicable to gene for gene relationships.

Gene-for-Gene Hypothesis

The concept of gene-for-gene hypothesis was first developed by Flor in 1956 based on his studies of host pathogen interaction in flax for rust caused by *Malampsora lini*. The gene for gene hypothesis states that for each gene controlling resistance in the host, there is a corresponding gene controlling pathogenesis in the pathogen. The resistance of host is governed by dominant genes and virulence of pathogen by recessive genes. The genotype of host and pathogen determine the disease reaction. When genes in host and pathogen match for all the loci, then only the host will show susceptible reaction. If some gene loci remain unmatched, the host will show resistant reaction. Now gene-for-gene relationship has been reported in several other crops like potato, sorghum, wheat, *etc.* The gene for gene hypothesis is also known as "Flor Hypothesis".

At molecular level, it is considered that gene for gene resistance usually involves production of toxins or antibiotic proteins by a resistant gene. The production of toxins is related to gene dosage. The resistance controlled by dominant gene is the most desirable. Gene relationships are rare or unknown for diseases caused by viruses, bacteria, *Fusarium* and organisms that cause rot.

A simple scheme to explain gene for gene relationship hypothesis (From Fehr, 1987)

Varieties	*Host genotype*	*Pathogen genotypes*	*Disease reaction*
1. One gene pairs	AA	Aa	Susceptible
	Aa		
	BB	Bb	Susceptible
	Bb		
	CC	Cc	Susceptible
	Cc		
2. Two gene pairs	AA BB	Aa or bb	Resistant
	Aa BB		
	Aa Bb		Susceptible
	AA CC	Cc	Resistant
	Aa CC	Aacc	Resistant
	Aa Cc		Susceptible
3. Three gene pairs	AA BB CC	aa bb	Resistant
	AA BB CC	aa cc	Resistant
	Aa Bb Cc	aabbcc	Susceptible

Mechanisms of Disease Resistance

Three different mechanisms operate in plants for disease resistance;, *viz.*, (1) establishment of the pathogen in the host tissue, (2) resistance to the growth and development the pathogen already established in the host tissue (3) ability of a host to perform well despite the establishment of the pathogen in the host tissues. The first two mechanisms considered as true forms of resistance and the last is termed as tolerance.

True Resistance

Vertical resistance is associated with hypersensitivity of the host cells leading to the cell death or chlorosis around the point of infection and consequent starving of the obligate parasite. This delays start of an epidemic and contributes to a defiance mechanism against specific race of a pathogen. The host prevents the establishment of certain races but not other races. In case of general resistance, the host plant resists to the development of all races of a pathogen. Disease resistance is governed by a number of morphological, physiological and biochemical features of host plant. Various morphological characters such as small, few sunken and hairy stomata with lesser opening duration and lesser opening duration of flower, have been found to be associated with disease resistance in different crop plants. These characters confer disease resistance by checking the entry of pathogen into the host tissues. Physiological characters like high osmotic pressure and acidity of cell sap and high crude fiber content have been reported to confers disease resistance. Biochemical studies revealed that high contents of tannin, proto catechuic acid, and catachol, phenol, alkaloids, silica and riboflavin were associated with disease resistance in different crop plants. Most of these chemical compounds are toxic to and thus check the establishment of the parasites into the host tissue.

Tolerance

The ability of a host to reproduce well despite the establishment of a pathogen in the host tissue is referred to as tolerance. The performance of tolerant plants is generally similar to resistant plants. However, such plants fail to check the entry and establishment of the host cell. Tolerant plants are always put in the category of susceptible plants because they exhibit the symptoms of disease infestation.

Nature of Resistance to Disease in some Crop Plant

Crop Species	*Disease*	*Cause of Resistance*
Potato	Late blight	R_1 to R_6 genes
Sugarbeet	Cerospora leaf spot	Saponin in leaves
		High poly phenol 3 hydroxy tyramin in leaves
		High dihydroxy phenylanine
		Low L-glutamic acid
Tomato	Leaf mould	Low sugar and free amino acid
	Fusarium and bacterial wilts	High alpha tomatin in root leaves and stem

Genetics of Resistance

There are three different modes of inheritance, *viz.,* (1) oligogenic, (2) polygenic (3) cytoplasmic. The inheritance of disease resistance is governed by these three modes.

1. Oligogenic Resistance

When the resistance is governed by one or few major genes, it is known as oligogenic resistance. In such cases, each gene has large and easily identifiable effect on resistance. The resistance may be governed by dominant or recessive genes. There are clearcut differences between resistant and susceptible plants. When the resistance is governed by single major gene, it is known as monogenic resistance. Oligogenic resistance is race specific. It provides resistance against one or few races of a pathogen. Such resistance involves hypersensitive reaction to the pathogen. A resistance governed by two or more major genes is generally more durable than that controlled by single major gene.

There are several examples of oligogenic resistance of disease resistance. Example of oligogenic resistance to disease include, in potato wart, potato virus and potato virus y. In some cases, oligogenic resistance is affected by some modifier genes. Transfer of oligogenic from one host to other is simple.

2. Polygenic Resistance

In some cases, the disease resistance is governed by several genes each having small and generally additive effect. The effect of each gene is not easily detectable. This type of resistance is referred to as polygenic resistance or quantitative resistance.

Main features of polygenic resistance are given below:

- ✰ It is non-race specific and therefore, provides protection from all the races of a pathogen.
- ✰ It is not possible to classify plants into different clear cut classes due to continuous variation for resistance.
- ✰ This type of resistance is generally more durable than oligogenic resistance because it involves several features of the host plant.
- ✰ Transfer of polygenic resistance from one host to another is more difficult than oligogenic resistance.

3. Cytoplasmic Resistance

Sometimes, disease resistance is governed by cytoplasmic genes. Examples of cytoplasmic disease resistance are resistance to yellow virus in sugar beet and resistance to leaf rust in wheat.

Sources of Resistance

In crop plants, there are four important sources of disease resistance. These are: (1) cultivated varieties, (2) germplasm collections (3) wild species (4) induced mutations.

1. Cultivated Varieties

In some crops, resistance to disease may be found in cultivated varieties. Resistant plants to curly top in sugarbeet and to mildew and leaf spot in alfa-alfa have been isolated from the commercial variety of respective crop. Cultivated varieties are the best sources of disease resistance, because they possess good agronomic characters besides disease resistance.

2. Germplasm Collections

Germplasm collections are the potential sources of disease and insect resistance in all the crops. In cotton, several germplasm lines resistant to bacterial blight and fusarium wilt have been identified based on screening of large number of germplasm in India. Generally, germplasm lines have poor agronomic characters. Hence, their use in breeding programmes poses some problems.

3. Wild Species

Related wild species are also potential sources of disease resistance. However, utilization of wild sources poses many problems such as cross incompatibility, hybrid in viability, hybrid sterility and linkages of several undesirable traits with desirable ones. Therefore, wild related species are only used as source of resistance when the desired resistance is not found within the cultivated species. Wild species of crops like potato, tomato, sugarbeet, *etc.*

4. Mutations

Both spontaneous and induced mutations are good sources of disease resistance. Disease resistance has been achieved in several crops through the use of induced mutations.

Sources of Disease Resistance in some Crop Plants

Crop Species	*Source of Resistance*	*Resistance for Disease*
Potato	*Solanum demissum*	Potato virus x, late blight and leaf roll virus
	S. spegazzinii	Leaf roll virus
	S. brachycarpum	Late blight, leaf roll virus, potato virus y
	S. brachycarpum	Leaf roll virus
	S. rybinni	Potato virus x
	S. acaule	Leaf roll virus, Potato virus x
	S. chacoense	Leaf roll virus, Potato virus x
	S. microdontum	Leaf roll virus
Tomato	*Solanum pimpinellitolium*	Fusarium wilt, leaf mould, curly top, tomato mosaic and spotted wilt virus
	S. minutum	Leaf mould
	S. esculentum var. minor	Leaf mould
	S. hirsutum var. glabratum	Leaf mould, bacterial canker, tomato mosaic
	S. peruvianum	Leaf mould, spotted wilt virus
	S. peruvianum var. dentatum	Curly top
	S. chinense	Canker, tomato mosaic
Sugarbeet	*Beta patellaris*	Cercospora leaf spot and curly top virus
	B. procumbens	
	B. artiplicifolia	
	B. lomatogona	
	B. vulgaris spp. Maritime	
Okra	*Abelmoschus manihot*	Yellow vein mosaic virus

Breeding Methods

Breeding methods for disease and insect resistance are the same as for other agronomic characters. Four breeding methods, *viz.*, (i) introduction, (ii) selection, (iii) hybridization (iv) mutation. The choice of resistance and mode of pollination of the crop species decides breeding methods.

1. Introduction

This is easy and rapid method of developing disease resistant variety. The resistant variety may be introduced and after testing, if found suitable, can be released in the disease prone area. In 1860, the grape crop in France was completely destroyed by the attack of *Phylloxera vertifolia.* Introduction of resistant root stocks to this pest from USA saved the grape crop from extinction in France.

2. Selection

When the source of resistance is a cultivated variety, mass selection and pure lines selection in self-pollinated crops, mass and progeny selections in cross-pollinated species and clonal election in the vegetatively propagated crops will be ideal for isolating disease resistant plants. The resistant plants may be multiplied, screened for disease resistance and released as a variety.

3. Hybridization

Hybridization is used when resistant genes are available either in the germplasm or in wild crop plants. After hybridization, the hybrid material is handled either by pedigree method or by backcross method. The pedigree method is used when the resistance is governed by polygenes and the resistant variety is an adapted one which also contributes some desirable traits. The backcross method is used when the resistant parent is un-adapted type or the resistance gene is to be transferred from wild species.

Durable Resistance

The resistance that lasts for a longer period is referred to as durable resistance. Durable resistance is of prime importance in the development of improved crop cultivars. Examples of durable disease resistance are in cucumber to *Cercospora* leaf spot; in potato to wart and potato viruses x and y; wild fir and angular leaf spot apple to scab and in cabbage, tomato and pea to fusarium wilt. The durability of resistance mainly depends on four factors, (1) formation of new race of pathogen, (2) genetics of resistance, (3) morphological features of host plant (4) bio associated with disease resistance.

Exploitation of Vertical Resistance

Vertical resistance refers to resistance of a host to specific race of a pathogen. The breeder should make best use of available genes for specific resistance to prevalent races. There are four different ways to exploit vertical resistance in plant breeding. These are: (1) development of cultivars with individual major genes, (2) development of multilines, (3) development of cultivars with major genes or gene pyramiding (4) gene deployment.

1. Varieties with Individual Major Gene

In this method an individual major gene is transferred for the prevalent race of pest into well adapted cultivar through backcrossing. Resistant plants are selected from segregating populations and backcrossed to the recurrent (adapted) parent. Generally, 5-6 backcrosses are sufficient to retain the genotype of recurrent parent with added resistance to specific race of a pathogen. This approach is widely used to control the prevalent race of a pathogen. Transfer of major genes from one host to another is very simple. The main disadvantage of this approach is that major gene may sometimes become susceptible to the minor prevalent races.

2. Development of Multiline

Multiline refers to seed mixture of isolines, related lines or unrelated lines. The seeds of genotypes with individual major genes are mixed together. This approach is used in self-pollinated species. The resistant genes to different races of a pathogen are transferred into different genotypes by separate backcross programme. Suppose there are five major genes each for a different race of a pathogen. These five major genes will be transferred in five different backgrounds. At the end seed of five resistant genotypes thus developed will be mixed together to constitute multiline. Thus multiline provides protection against several races of a pathogen. The new race of a pathogen whenever occurs will attack few genotypes of a multiline. All the component genotypes of a multiline will never become susceptible to new race of a pathogen. The resistant components in a multiline will act as a barrier and delay the spread of the new race to susceptible plants. There are two main disadvantages of this method. First, several backcross programmes have to be launched to transfer each major resistant gene in a separate background. Second, the recurrent parent limits the agronomic performance of the multiline. Moreover, it requires extensive testing against specific races.

3. Gene Pyramiding

Gene pyramiding refers to incorporation of two or more major genes in the host for specific resistance to a pathogen. This it differs from multiline, where each major gene is incorporated into separate genotype and then resistant genotypes are mixed together. In pyramiding approach major genes are incorporated in a single cultivar. Combination of several major genes for specific resistance provides protection against several new races that may develop in the pathogen. The new varieties can be developed with a combination of 2, 3 or more major genes such in Canada. The main demerits of this technique are given below:

- Lot of efforts have to be made to incorporate several major genes into single cultivar. Gene pyramiding approach has been used in oat against crown rust.
- It requires extensive testing against various races of a pathogen to ensure transfer of desirable alleles for resistance.
- If the same genes are used singly in other cultivars, pyramiding approach will enhance the chances of formation of new virulent races of pathogen.
- The agronomic performance of the new cultivar is restricted to the recurrent parent when backcross is used.

4. Gene Deployment

Another way of exploiting vertical resistance is gene deployment. Gene deployment refers to planned or strategic use of major genes in development of resistant cultivars for various geographical areas. This approach helps in preventing diseases and maintaining the diversity for major genes in different geographical regions. This approach requires a number of genes with similar effectiveness for control of prevalent races. However, it is impossible to find out different genes with similar effectiveness. This is the main drawback of gene deployment approach. This approach was used for soybean breeding in Northern United States, but was not found effective.

Screening Techniques

Screening for disease resistance is carried out under field and glasshouse conditions. The glasshouse tests are conducted under controlled conditions and, therefore, glasshouse screening is considered more reliable than field tests. However, field and glass house screening are equally important. The procedure of field inoculation differs for various types of diseases.

1. Soil-Borne Diseases

For soil-borne diseases like root rots, collar rots, wilts, *etc.*, the screening is done in disease sick Plots. Disease sick plots are developed in three ways: (1) by mixing soil from other sick plots, (2) by adding remains of diseased plants (3) by adding inoculum developed in the laboratory. The soil from disease sick plots can be used in pots for conducting tests in the glass house.

2. Air-Borne Diseases

For air-borne diseases such as rusts, smuts, mildews, blights, leaf spots, *etc.*, the screening is done either by dusting the spores or by spraying the spore suspension on the healthy plants. Planting of highly susceptible varieties after few rows of test material is also used to develop the inoculum.

3. Seed-Borne Diseases

In case of seed-borne diseases like smuts and bunts, either dry spores are dusted on the seeds or the seeds are soaked in the spore's suspension.

4. Insect Transmitted Diseases

In case of insect transmitted diseases, the insect from susceptible varieties are collected and released on healthy plants or juice of diseased plants is rubbed onto healthy plants after causing mechanical injury in healthy plants.

Practical Achievements

Disease resistant varieties have been developed in many crops all over the world. In India disease resistant varieties have been evolved in many horticultural crops. Almost all the currently released varieties in okra, a yellow mosaic virus resistant variety Parbhani Kranti has been released. Now several hybrids of okra having resistance/high level of tolerance to yellow vein mosaic virus have been commercialized by the private seed companies in India.

Chapter 9

Breeding for Insect Resistance

Introduction

Like diseases, insects are important causal factors of biotic stress in crop plants. Insects attack all the crop plants and lead to considerable losses in yield as well as quality. Insect attack leads to various types of damages in crop plants such as (i) Reduction in plant growth or stunting (ii) Damage of vegetative (stem leaves and branches) and reproductive (flower bud, flower, fruit and seeds) parts, (iii) Premature defoliation (iv) Wilting of plants. Insects causes 14 per cent estimated yield loss of all important crops on global bases. Insects cause yield loss directly either by sucking cell sap (sucking insects) or by eating away various plant parts (tissue feeding insects). Insects also cause yield loss through transmission of various diseases.

There are two important methods of insect control, *viz.,* (1) biological method (2) chemical method. In biological method, insects are controlled in three ways, *viz.,* (i) by the use of predators and parasites (natural enemies) of insect pests, (ii) by using botanical pesticides such as *neem, datura, ipomea* in the form of leaf extracts (iii) use of resistant varieties. Biological method is cheap and does not have any adverse effect on the ecosystem, though lesser effective than chemical method of insect control. The chemical method includes use of various chemical insecticides. Use of insecticides has several disadvantages. It increases cost of cultivation, reduces population of predators and parasites of insect pests, leads to environmental pollution and development of pesticide resistant biotypes of insects. Thus genetic resistance is the cheapest and the best method of insect control in crop plants. Genetic resistance refers to the ability of some genotypes to give higher yields of good quality than susceptible varieties at the same initial level of insect attack under similar environmental conditions. Thus resistance is defined in relation to susceptible varieties. This chapter deals with breeding for resistance to insect pests and parasitic weeds.

Mechanisms of Insect Resistance

There are four mechanisms of insect resistance, *viz.,* (1) non-preference, (2) antibiosis, (3) tolerance (4) avoidance or escape. The first three mechanisms were given by Painter (1951) and the fourth one was added subsequently. A resistant variety may have one, two or more mechanisms.

1. Non-preference

Non-preference refers to various features of host plant that make the host undesirable unattractive to insects for food, shelter or reproduction. This type of insect resistance is also known as non-acceptance and antixenosis. Non-acceptance appears to be more because in most known examples of this type of resistance, insects will not accept plant even if there is no alternative source of food. Various plant characters which non-preference includes are colour, light penetration, hairiness, leaf angle, odour and taste. For example, in okra leaf, open canopy, nectarilessness, fragrant, bract, thickness and hardness of boll rind and long pedicel are examples of non-preference to bollworms and hairiness of leaf and stem is non-preference for jassids. In pea, yellow green genotypes are less preferred by pea aphid than blue green genotypes. In soybean, hairy genotypes are less preferred by potato leaf hopper than smooth types. In potato, resistance to Colorado potato beetle is related to odour.

Non-Preference Mechanism of Insect Resistance in some Crop Plants

Host Crop	*Insect Pest*	*Host Plant Character*	
		Non-preference	*Preference*
Pea	Aphid	Yellow green leaves	Blue green leaves
Cabbage	Aphid	Leaves with high light reflection	Leaves with low light reflection
Sugarbeet	Aphid	Low free sugar	High free sugar

The degree of non-preference varies from species to species. In some cases, non-preference is so strong that insects migrate from resistant plants, for example aphid resistance in raspberry. In other cases, insects do not walk away from resistant plant. They feed for shorter period but feel restless than on susceptible plants. The example is aphid resistance in sugarbeet.

2. Antibiosis

Antibiosis refers to the adverse effect of host plant on the development and reproduction of insect pests which feed on resistant plant. Resistant plants retard the growth and rate of reproduction pests. In some cases, antibiosis may lead even to death of an insect. Antibiosis is considered as the true form of resistance to insect pests. Antibiosis may involve morphological physiological and biochemical features of the host plant.

3. Tolerance

Tolerance refers to the ability of a variety to produce greater yield than susceptible variety at the same level of insect attack. In other words, a tolerant variety will give higher yield than susceptible one despite the insect attack. The tolerance is measured in terms of rejuvenation potential (capacity to repair and replace damaged parts), healthy leaf growth, flowering compensation potential and superior plant vigour. Tolerant cultivars have greater recovery of damaged parts than susceptible ones.

Antibiosis Mechanism of Insect Resistance in some Crop Plants

Host Crop	*Insect Pest*	*Cause of Antibiosis*
Sugarbeet	Aphid	Low free sugar
Potato	Aphid	Gummy trichome exudates

4. Avoidance or Escape

Avoidance refers to escape of a variety from insect attack either due to earliness or its cultivation in the season where insect population is very low. Avoidance is also an effective means of protecting crop from the damage of insect pests.

Basis of Insect Resistance

There are the three important bases of insect resistance, (i) morphological, (ii) physiological (iii) biochemical features of host plant. In other words, insect resistance is mainly governed by morphological, physiological and biochemical features of host plant. These are briefly discussed as below:

1. Morphological Factors

Various morphological traits, *viz.*, hairiness, colour, thickness and toughness of tissues, and several other characters are known to confer insect resistance in different crop plants.

- **Hairiness-** Hairiness of leaves is associated with resistance to many insect pests. For example, resistance to aphid in turnip and leaf beetle in cereals is associated with hairiness of leaves and stem.
- **Colour of plant-** Plant colour contributes to non-preference in some cases. For example, red cabbage and red leaved Brussel's sprouts are less favored than green types by butter flies and other *Lepidoptera* for ovipositor.
- **Solid stem-** In wheat, solid stem confers resistance to stem sawfly.
- **Toughness of tissues-** Tough and thick plant tissues cause mechanical obstruction to feeding and oviposition insect pests.

2. Physiological Factors

Some physiological factors such as osmotic concentration of cell sap and leaf exudates are associated with insect resistance. Some species of *Solanum* secrete gummy exudates from hairs the leaves. Aphids and Colorado beetles get trapped in such exudates and are unable to feed produce. Several species of *Solanum* and some varieties of tobacco, secrete exudates from glandular leaf hairs which are toxic to various insects and mite pests. In *Medicago disciformis* secondary trichomes on the leaves secrete antibiotic exudates. These exudates kill the alfa-alfa weevil at higher concentrations and retard the growth at lower concentrations.

3. Biochemical Factors

Biochemical factors are considered as more important than morphological and physiological factors for insect resistance. A number of biochemical substances are known to be associated with insect resistance in different crop plants. Some examples are leaves of wild tomato (*Solanum hirsutum* var. *glabratiam*) contain highly active ethanol soluble compound which is lethal to tomato fruit worm and tobacco flea beetle.

Certain bio-chemical substances act as feeding stimuli for insect pests. Lack or low concentration of such substances in host plant will lead to non-preference type of resistance. For example in Brassica, sinigrin (oil glucoside) act as feeding stimulant to cabbage aphid. Low concentration of sinigrin in the leaves confers resistance to this insect pest in Brassica. In sweet clover (*Melilotus* spp.), common in provides feeding stimulus to flying weevil.

Genetics of Insect Resistance

The inheritancence of insect resistance may be governed in three ways, (1) oligogenes, (2) polygenes (3) cytoplasmic genes. Examples of oligogenic insect resistance include in alfa-alfa to pea aphid, in raspberry to rubus aphid, in apple to woolly aphid, *etc.* The oligogenic resistance is also affected by some modifier genes in some cases. The transfer of oligogenic resistance from one host to another is simple.

In some cases, insect resistance is governed by several genes each gene having small and additive effect. Examples of polygenic insect resistance are: in brinjal to stem borer, in alfalfa to spotted aphid, *etc.* Sometimes, insect resistance is governed by cytoplasm genes. Examples of insect resistance governed by cytoplasm genes include resistance to root aphid in lettuce, more such cases may be observed if extensive research is carried out.

Sources of Insect Resistance

There are five different sources of insect resistance in crop plants, (1) cultivated varieties (2) germplasm collections (3) wild species (4) mutations (5) micro-organism, briefly discussed below:

1. Cultivated Varieties

In some crops, genes for insect resistance may be found in cultivated varieties. The insect reactions of cultivated varieties are known in almost all the crops. The desirable source of resistance can be selected from the cultivated varieties. Cultivated varieties are the best sources of resistance, because they have good agronomic characters, besides resistance.

2. Germplasm Collections

Germplasm or genetic resources are good sources of insect resistance. Resistant lines are identified by screening of germplasm for specific insect. In apple, 14 lines resistant to rosy aphid and 3 lines immune to green apple aphid were identified through screening of 2000 apple germplasm lines. Many such examples of insect resistance can be cited from other crops.

3. Wild Species

In several crops, resistant genes for insect are found in the wild species or wild relatives of crop plant. Wild species are good sources of insect resistance. For example, resistance to Rubus aphid of raspberry is found in wild species of raspberry. In tobacco, resistance to root knot nematode is obtained from wild species.

4. Induced Mutation

Sometimes, insect resistance is obtained through induced mutations. Insect resistance has been obtained in many crops by this method

5. Micro-organisms

Now micro-organisms are being used as source of resistance to insect pests. In USA, Monsanto Company has transferred a gene from *Bascillus thuringiensis* (Bt.) into the system of cotton plant through genetic engineering. The Bt. gene is believed to provide effective resistance against bollworms. When the bollworm larva punctures the boll, a toxin is secreted by the plant which leads to death of the larvae by a slow process. Examples are large number of cotton and brinjal hybrids released in India using this Bt. Gene. These cotton hybrids are called as Bollgard I and Bollgard and these are very popular in India, occupying almost 100 per cent of the total cotton area in India.

Durable Resistance

It refers to long lasting resistance. Examples of durable insect resistance are: in raspberry to aphid, in apple to wooly aphid and in alfa-alfa to spotted aphid and pea aphid.

The durability of resistance depends mainly on four factors (i) formation of new races/biotypes, (ii) genetics of resistance, (iii) morphological features of host plant (iv) bio-chemical substances associated with resistance.

1. New Biotypes

The formation of new insect biotypes is much lower than physiological new races of resistance. The formation of new races of fungal and bacterial pathogens is very common. New races are frequently observed for rusts and mildew. New physiological races are formed through spontaneous mutations and hybridization.

2. Genetics of Resistance

The mode of inheritance of resistance also affects the durability of resistance. In general, monogenic resistance is lesser durable than oligogenic and polygenic resistance. However, in some cases, monogenic resistance is very durable. For example, resistance to stem eelworm controlled by single dominant gene is very durable. On the other hand, in some cases, polygenic resistances overcome by new biotypes. For example, polygenic resistance of rapeseed varieties to aphid overcome by new resistance breaking biotypes of cabbage aphid in New Zealand and England The polygenic resistance involves several resistance mechanisms which lead to longer durability.

3. Morphological Characters

Insect resistance associated with morphological and anatomical features of host plant is more durable than other kinds of resistance. These morphological features act as non-preference as well as antibiosis for the insect. Cases of durable disease resistance associated with morphological features of host plant are less known.

4. Biochemical Factors

In some cases, durable resistance is associated with biochemical substances present in the host plant. For example, durable resistance to spotted and pea aphid in alfa-alfa with high siphoning content.

Breeding Methods

Breeding methods for developing insect resistant varieties are the same as for other economic or agronomic characters. Five breeding methods, (1) introduction, (2) selection (pure-line and mass selection depending upon mode of pollination) (3) hybridization (4) mutation (5) transgenic breeding are used for development of insect resistant varieties. The choice of breeding methods depends on source of resistance, genetics of resistance and of the crop species.

Screening Techniques

Screening for insect resistance is done under field and glasshouse conditions. In field larger number of plants can be screened than in glasshouse. Moreover, the material is also exposed to other prevalent insect pests of the area. However, in field screening it is not possible to ensure uniform initial infestation of all the plants in a population. The following techniques are used to promote uniform infestation by an insect pest in the field:

- Inter-planting one row of susceptible variety between two rows of the test material.
- Screening of the plant material in the insect prone areas.
- Screening of material in the season when there is heavy infestation of an insect pest. For example, there is more infestation of stem borer in rice in off season than in main season.
- Transferring equal number of eggs to each plant by hand.

In glasshouse smaller number of plants can be screened than in the field. However, results of glasshouse tests are more reliable than field tests. Glasshouse tests have been conducted in rice to stem borer and in alfa-alfa to spotted alfa-alfa aphid. Both glasshouse and field tests are important. It is advisable to go for field screening after glasshouse tests. The material found resistant in glasshouse should also be tested under field conditions.

Practical Achievements

Insect resistant varieties have been developed in many crops. Phylloxera resistant varieties of grapes have been developed in France through the use of resistant root stocks introduced from USA. In USA Hessian fly and stem sawfly were the serious insect pests of wheat. More than 23 Hessian fly resistant varieties have been developed. These varieties are successfully grown in Hessian fly prone areas of USA. Stem sawfly resistant variety Rescue' with solid stem was released from Canada in 1946. This variety is successfully grown in stem sawfly prone areas in USA and Canada. In maize, resistant varieties to rust have been released and maize rust is no longer a serious disease in USA. A barley variety 'Will' resistant to green bugs has been developed in USA. Cultivation of this variety has reduced the population of greenbugs to 50 per cent. Alfa-alfa varieties "Cody', Moapa and Zia, developed in USA, are resistant to spotted alfa-alfa variety.

Insect resistant varieties have been developing in many other crops. For example, in raspberry to aphid, in lettuce to root aphid, in apple to woolly apple aphid, in grapes to *plnylloxera, etc.*

Merits and Demerits of Resistance Breeding

There are several merits and demerits of breeding for insect and disease resistance. Some merits and demerits of resistance breeding are given below:

Marits

- Resistant varieties play an important role in controlling the losses caused by diseases and insects in crop plants. About 14 per cent annual losses of potential crop yield are caused by insect pests and about 20-30 per cent losses are caused by various diseases. These losses can be minimized through the use of resistant varieties.

- ✰ Resistant varieties lead to reduction in the cost of production resulting in increasing the cost benefit ratio. Use of resistant varieties leads to reduction in the use of pesticides which in turn results in reduction of environmental pollution and health hazards caused by the use of pesticides. Moreover, genetic resistance protects natural enemies of insect pests (Predators and parasites) which are killed through the use of insecticides.
- ✰ Resistant varieties are non-toxic to man, farm animals and wild life. In other words resistant varieties do not contain pesticide residues which are toxic to man, farm animals and wildlife.
- ✰ Genetic resistance is the only solution of some diseases such as wilts, rusts, smuts, nematodes and bacterial blights.

Demerits

- ✰ It is a long-term process which takes 10-15 years to develop agronomically accepts variety even when the sources of resistance are readily available.
- ✰ In some cases, breeding for resistance to one pest leads to the susceptibility to another pest. This is because the host plant feature associated with resistance to one insect is associated with susceptibility to another insect.
- ✰ In many cases, genes for disease and insect resistance are available only in the related wild species. Interspecific gene transfer poses many problems. Moreover, resistant genes are associated with some undesirable characters. It takes a long time to discard undesirable genes in a breeding programme.
- ✰ Breeding for disease and insect resistance is an expensive method which requires adequate funding for a long period being a long-term process. Often lack of funds leads to the interruption of such valuable programmes.
- ✰ In some cases, the resistant variety has lower yield and poor quality. For example, 'Rescue' a stem sawfly resistant variety of wheat, developed in USA has lower yield and poor grain quality than sawfly susceptible varieties in the absence of sawfly incidence. In such situation a crop rotation of resistant and high yielding susceptible variety is adopted. Two years rescue-three years high yielding susceptible variety rotation will keep the sawfly to an acceptable level and help in realization of good wheat yield.

Chapter 10

Intellectual Property Rights

Introduction

The property refers to wealth or valuable things earned by a person. It is estimated in terms of land, house, garden, industries, animals, gold, silver, diamond, money, *etc.* accumulated by a person. The ability of the mind to come to correct conclusion about what is true or real and about how to solve problems depends upon the learning the things around and the mental ability of person. Common types of IPR include copyrights, trademarks patents, industrial design right plant varieties and farmer right (PPV and FR) and trade secrets.

Types of Property

It is classified in two groups:

- **Immovable property:** It refers to fixed type of properties which cannot move from one place to other place, *e.g.,* Land, building and gardens.
- **Movable property:** It refers to the property which can be easily sifted from one place to other place, *e.g.,* Animals, farm machines, furniture, fixtures, gold, silver, diamond and money.

Intellectual Property

There is third type of property what we call intellectual property. The product/process/idea which is outcome of the brain of a person and can be used on commercial scale for benefits of human kind are called intellectual property. In other word intellectual property refers to creation of the mind inventions, literary and artistic work and symbols names, images and design used on commercial scale. The intellectual property is an old phenomenon as our evaluation as human.

The protection of intellectual property started with increasing environmental and social competition. To live on earth within its limited natural resources human started to select things on the basis of his intellect and innovative thinking. Each crop plant is considered to have originated in a specific region of the World. But the modern era of intellectual property started in India since 1856 in case of patent by granting some exclusive right to inventor initially for 14 years. Again the act was re-enacted after some minor modification in year 1859.

Main Features of Intellectual Property

1. It is measured in terms of new ideas, processes products, inventions and innovations developed by a person.
2. It requires lot of intellectual inputs in terms of thinking planning and fine tuning of new ideas/product/processes, *etc.*
3. It requires considerable amount of funds and other resources to developed new products/process.
4. The main problem with intellectual property is that it can be copied reproduced and used by others resulting in loss to inventor unless protected by some laws/rules.

Categories or Form of IPR

It is broadly divided in two groups:

(A) Primary Rights

The TRIPS Agreement of WTO recognizes seven categories of IPR, which were governed by several WIPO administered treaties:

- Copyright and related rights.
- Trademarks, trade name and service marks.
- Geographical indications.
- Industrial design rights.
- Patents.
- Layout designs of integrated circuits.
- Undisclosed information.

(B) Sui-generis Rights

It refers to things of their own kind or things with unique characteristics. Such rights include database right, mask work, plant breeder rights, traditional knowledge, moral right and supplementary protection certificate.

(A) Primary Rights

(1) Copyrights and Related Rights

Copyrights refers to a document which grants exclusive right to the author/ creator to publish and sell literary or musical or artistic work. The right of an author; artist, publisher, *etc.* to retain ownership of work and produce or contract others to produce copies is called copyright.

Copyright is the most commonly used low of protections of all intellectual property concerned with the work of human intellect. The aim of copyright is the protection of literary and artistic works. These include writhing, music and works of the fine arts, such as painting and sculptures, and technology based works such as computer programs and electronic databases. Copyright protects the expression of thought and not ideas. The term of protection for copyright in the Berne convention the minimum requirement is 50 years after the death of author and for photographic works and works of applied art the minimum term of protection is 25 years after the making of the work.

Main features of Copyrights

1. It provides protection for a specific duration.
2. It is application in all countries.
3. The copyright holder has the rights to authorize others to use the protected work.
4. This is applicable to books, movies, music, painting, photograph and software.
5. Related right used in opposition to the term authors rights.
6. Related right is also known as neighbouring rights.

(2) Trademark, Trade Names and Service Marks

As long as 3000 years ago, Indian crafts men used to engrave their signature to their artistic creation before sending them to Iran. Trademark may consist of word, designs, letters, numerals or packaging, slogans, devices, devices, symbol, *etc.* It usually ensures a consistent level of quality be it good or bad. A trade names also known as a trading name or a business names, is the name which a business trade under for commercial purpose. Pharmaceutically also have trade name (*e.g.,* Aspirin) after dissimilar to their chemical name (acetylsalicylic acid). The commercial name by which a chemical is known as called trade name.

Main features of Trade Mark and Trade Name

1. It can be a word, name, symbol and device or mark which is used to identify and distinguish the goods or services of one company from goods or services of another.

2. It is used to identify its product and to distinguish them from others. It is name particular of company.
3. Trade name may or may not be registered.
4. Trade name is exclusive or no exclusive.
5. Trade name is used on letter heads and bank accounts.
6. Trade name is business name of person or organization making.
7. "General Motors Inc" is the trade name of company, Kleenex is the trade name for a brand tissue, xerox a single brand of copier.

(3) Geographical Indication

The world intellectual property organization (WIPO) helps in protection of geographical indication (GI). Geographical indications have been registered by Lisbon Agreement members up to.

"A geographical indication is a sign used on goods that have a specific geographical origin and often processes qualities or a reputation that are due to that place of origin (town, region country). Geographical indication is the name of region, a specific place or in exceptional cases a country used to describe an agricultural, natural or to manufactured goods product or a food stuff. Geographical indications are notice stating that a given product originates in a given geographical area.

For example- Banaras brocades and sarees (Varanasi-UP), Handmade carpet (Bhadohi-UP), Lucknow chikan craft (Lucknow-UP), Malihabad Dashehari mango (Lucknow-UP), Darjeeling tea (AS), Kangra tea (AS), Mahabaleshwer strawberry (MH), Sangali turmeric (MH), Sahi litchi (Bihar), Alphanso mango (MH), *etc.*

(4) Industrial Design

Industrial design is that aspect of a useful article which is ornamental or aesthetic. It may consist three dimensional features such as a shape or surface of article or two-dimensional such as patterns lines or colour. Registration and renewals provides protection for in most cases up to 15-20 year. The TRIPS provides for protection of a minimum of 10 years.

Main features of Industrial Design

1. It is an applied art where by the aesthetic and usually of products may be improved for marketability and production. Industrial design often utilizes 3D.
2. Industrial design refers to any original shape, picture or some combination applied to a useful article of manufacture.
3. Industrial design refers to the profession service of creating and developing concepts.

(5) Patents

Patents are granted as an exclusive right by the government for an invention for a limited period of time in consideration of disclosure of the invention by the inventor or his representative. An invention is a new product or process involving an inventive step and capable of industrial application. Patent refers to a document granting and inventor sole rights to an invention.

Main features of patent

1. It is a legal document issued by a federal government that grants exclusive right for the production sale and profit from the invention of a product or process for a specific period of time.
2. Patents also grant the right to prevent others from copying the invention.
3. An official license is granted by the patent office to issue right to an individual.
4. It is a document that allows the patent owner to prevent others from making using or selling.

(6) Layout Design of Integrated Circuits

Layout design (topography) of integrated circuits appeared with computer technology and has acquired importance as the technology makes rapid advances. The term layout design is used for layout of transistors and other circuitry elements and includes leads wires connecting such elements and expresses in any manner in a semiconductor integrated circuit.

(7) Undisclosed Information

Undisclosed information or trade secret are any business information on a new manufacturing method, know-how, chemical formulae, pattern, idea, blue-print or prototypes, sales, methods, distribution methods. A trade secret is anything (a formula, process, method, mechanism, tool, pattern or device) which is disclosing party desires to keep secret.

Main features of Trade Secrets

1. There is no specific period for trade secrets. It may continues lifelong or for generations together.
2. There is no need of registration for trade secrets.
3. It does not provide opportunity to others for improvement of innovation.
4. It is not applicable to books, equipments, plant varieties, design which are openly used.

(8) Domain Names

It refers to the unique name that identifies as internet site. Domain names have two or more parts separated by dots, *e.g.*, www.microsoft.com, It is without knowing the true numerical address, web address.

(B) Sui-generis Rights

Sui-generis is Latin expression, literally meaning of its own kind/genus or unique in its characteristics.

- **Data base right-** Given to computer database duration last for 15 years.
- **Mask work-** It is a two or three dimensional layout of an IC (IC or chip).
- **Plant breeder rights-** It is also known as plant variety right for new variety. Duration 15 years for field, 18 years for tree, fruit, ornamental plants. PPV and FRA-2001 requires novelty, distinctiveness, uniformity, and stability (N-DUS).
- **Farmer rights-** It is legal Right provided to farmers to save, use sow, replant, exchange, and share or sell his form produce under PPV and FRA-2001.
- **Moral rights-** It is a spatial extension to copyright that gives the originator of a copyright work right over its use.
- **Supplementary protection certificate-** It is patent like, *e.g.,* Medicinal and plant protection products.
- **Indigenous intellectual property-** The knowledge, innovation, practices of India people and local communities.

Traditional Knowledge and Bio-piracy

Such as agricultural knowledge medicinal bio-diversity, *e.g.,* Charak Samhita, Susrutha, *etc.* medicinal "Used turmeric in wound healing" fungicidal uses of neem oil. The Texas Company was already trading in Basmati rice. Brand name- Texmati, Jusmati and Kasmati. Edible hearbal mixute, jamun, gurmar, bitter gourd and brinjal.

Requirement and Process of Protection for Plant Varieties and Breeders Right (N-DUS)

1. **Novel:** If at the date of filing of the application for registration for protection the propagating or harvested material of such a variety has not been sold, in India earlier than one year, outside India in case of tree earlier than six years.
2. **Distinct:** If it is clearly distinguishable by at least one essential characteristic from any other variety whose existence is a matter of common knowledge in any country at the time of filing of the application.
3. **Uniform:** If subject to the variation that may be expected from the particular feature of its propagation, it should be sufficiently uniform in its essential characteristic.
4. **Stable:** If its essential characteristic remains unchanged after repeated propagation or in the case of a particular cycle of propagation at the end of each cycle.

Advantages

- ☆ It promotes healthy competition for invention/innovation among the intellectuals.
- ☆ It helps in improving the quality of the product.
- ☆ It makes available new ideas/technologies to different countries.
- ☆ It leads to faster development of industries in research and development work.

Disadvantages

- ☆ The procedure of registration, particularly of patents is very lengthy.
- ☆ It involves lot of money transaction in registration, renewal and licensing.
- ☆ It invites lot of court cases due to infringements.
- ☆ It may lead to monopoly of right holders, *etc.*

Section-II

Cytotaxonomy and Breeding Behaviour of Fruit Crops

Chapter 11

Mango

Botanical name : *Mangifera indica* L.

Family : Anacardiaceae

Chromosome number : 2n = 2x = 40

History of Mango Breeding

Mango is one of the choicest fruits of India, grown over an area of 1.23 million hectares in the country. Mango occupies the prime position in India as apple in temperate and grape in subtropical areas. In India, mango is acclaimed as 'King of fruits'. The name *Mangifera* was given for the first time by Bontius in 1658, when he referred to this plant as arbor *Mangifera* (the tree producing mango). Linnaeus also referred it as *Mangifera arbor* in 1747, prior to changing the name to its present form (*Mangifera indica*) in 1753. Mango is a good source of vitamin A and C apart from the usual content of minerals and other vitamins. Mango is also considered to have some medicinal properties. Ripe fruits of mango are fattening, diuretic and laxative. The kernel is effective against diarrhoea and asthma. Besides table purpose, fruits of mango can be used for the preparation of pickles, preserves, jam, amchur (mango powder) and mango leather (ampapad).

Botany

Flowering in mango is preceded by the differentiation of the flower bud in the shoots. Period of differentiation is reported to be October-December depending upon the climatic conditions. Mango inflorescences also arise from axillary buds quite frequently. Panicle is much branched having several hundred tiny flowers (1000-6000) which may be male and hermaphrodite (andromonoecious). The percentage of perfect flowers varies from 0.74 per cent to 70 per cent depending upon the varieties. Occurance of few abnormal flowers, *e.g.,* double hermaphrodite flowers with 10 sepals and 10 petals, 2 stamens and 2 ovaries. Pollination is essential for fruit set in mango. Pollination in mango is mainly entomophilous (house fly). It was found that natural pollination about 40-60 per cent perfect flowers failed to receive any pollen grains.

Germplasm Resources

India is the home of mango germplasm where more than thousand varieties are existing, which are widely distributed in different agroecological zones. Central Institute for Subtropical Horticulture, (CISH) Lucknow has the largest collection of mango (633 accessions in the national repository) and they have greater genetic variability with respect to fruit shape, skin colour, stone size, period and time of maturity, pulp thickness, colour, bearing habit, yield and quality parameters. Further, IIHR, Bangalore, IARI, Pusa, New Delhi, Sabour (Bihar), Fruit Research Station Sangareddy (Andhra Pradesh), *etc.* are also maintaining the germplasm of mango. In India, majority of varieties are monoembryonic whereas in most tropical region polyembryonic types are predominant.

Almost all the commercial cultivars of mango are related to a single species *Mangifera indica*. However, a few commercial cultivars of South East Asia belong to other edible species such as *M. altissima. M. caesia, M. cochinchinensis, M. foetida, M. griffithi, M. langinifera, M. longipes, M. macrocarpa, M. odorata, M. pajang, M. pentandra, M. sylvatica* and *M. zeylanica*. There are different reports regarding the number of species in Genus *Mangifera*. Singh (1969) reported 62 species whereas Mukherjee (1949) reported 41 species but later on he reported that only 39 species are existing (Mukherjee, 1985). There are five species of *Mangifera* reported from India, *e.g., M. andamanica, M. indica, M. khasiana, M. sylvatica* and *M. comptosperma* (Mukherjee, 1985).

Breeding Objectives

Qualities of an ideal mango variety have been outlined as follows:

- ☆ Dwarf tree growth habit.
- ☆ Precocity and regularity in bearing.
- ☆ Attractive and good quality fruits.
- ☆ High productivity and resistance to major diseases and pests.
- ☆ Good transport and processing qualities.

Breeding Methods and Achievements

Introduction

For incorporation of good colour to boost export of fresh fruits, a number of mango varieties were introduced from different countries for use as donor parent. Tommy, Ziulete, Haden, Sensation and Julie are the coloured varieties of mango which were introduced from Miami, Florida (USA). Other varieties, Pl 24927,

M 4336 (Carabao) from USA and EC 201556 (Carabao) from Phillippines were introduced as regular bearing varieties, cultivar Amolie and Sweet were introduced from Belgium and Thailand respectively.

Selection

Almost all the present commercial varieties of mango in the world were developed from open-pollinated seedling selection, *e.g.*, Dashehari, Langra, S.B.Chausa, Rataul, Swarnarekha, *etc.* The evolution of Florida varieties which are the leading mango cultivars of the world is interesting. In 1889, introductions were made from India of which Mulgoa became popular. Cultivar Haden was a seedling of Mulgoa, subsequently, many promising seedlings were selected which became popular. Tommy Atkins from Haden, Keitt from Mulgoa, Dyke and Palmer from unknown origin, Irwin from Lippins, Golden Nuggets and Brooks from Sandersha, Sensation from unknown origin, *etc.* are promising seedling selections.

Clonal Selection

Exploitation of natural variability for selection of superior clones of commercial mango cultivars has been undertaken. Clonal selection has also resulted in identification of few elite clones. Dashehari-51 from Dashehari, a regular bearer (CISH, Lucknow), 'Subash', a chance seedling from Zardalu (BAC, Sabour), Red blush, a strain of Alphonso (Vengurla), heavy yielding strains of Langra and Himsagar (Kalyani, W.B.), bacterial black spot resistant clones of Kensington, superior clones of Rumani and Neelum (Tamil Nadu) and a regular bearing cultivar 'Cardoz Mankhurad' in Maharashtra which is selected from Goa Mankurad. In Maharashtra, one off-season selection 'Niranjan" has been made at Parbhani, which comes to flowering during June to July and the fruits mature in October. In TNAU (Regional Research Station, Paiyur), a clonal selection from Neelum was identified as dwarf variety and released as Paiyur-1. This is suitable for high density planting (400 plants/ha).

Hybridization

In mango hybridization, work taken up in post-independence period laid emphasis on regular and precocious bearing, dwarfness, high percentage of pulp, fibreless flesh, large fruits with red blush, good keeping quality and freedom from spongy tissue. In recent years, emphasis has also been laid on evolving varieties tolerant to mango malformation. A variety Bhadauran, tolerant to this disorder, was developed through hybridization between Neelum and Dashehari. The work at Sabour yielded two promising hybrids namely Mahmud Bahar and Prabha Shankar from the parental combinations of Bombai x Kalapady. Hybrid Mahmud Bahar was found to be a regular bearer for four years whereas, Prabha Shankar was not a regular bearer. Further, the work on improvement of mango was initiated at Saharanpur in 1951 and also in Punjab in 1950 to develop regular bearing varieties. Later on, in India, nearly 20 inter-varietal hybrids of mango have been released for cultivation from IARI, New Delhi, CISH, Lucknow, IIHR, Bangalore, FRS, Sangareddy,

HC and RI, Periyakulam, AES, Paria (Gujarat), FRS, Vengurla, *etc.* of the hybrids developed in India, Mallika and Ratna have received commercial recognition. The cultivar 'Sindhu' evolved through intensive back crossing between Ratna and Alphonso develops fruits parthenocarpically under natural temperature conditions.

Inter-specific Hybridization

Inter-specific hybridization did not receive more attention but it can be a useful tool to transfer some useful genes in cultivated varieties. This is possible because all the *Mangifera* species have the same chromosome number (2n = 40). Therefore, they can inter cross easily.

Improved Hybridization Technique

a. Single day pollination of limited number of flowers in a panicle is the ideal practice. Here, the main emphasis was given on utilizing large number of panicles and crossing whatever few flowers opened on the panicle during that single day. Bagging with perforated polythene bags of 24" x 12" size of 100 gauges was preferred. Crossing of a few flowers in a given panicle at one time is advocated than taking up crossing in more number of flowers in a given panicle in batches over a number of days.

b. **Caging technique:** The discovery of self-incompatibility in some of the popular cultivars at IARI, New Delhi led to further improvement in the technique of hybridization. It is known as caging technique (Sharma and Singh, 1970, Singh *et al.,* 1962). In this technique, grafted plants of parent varieties are enclosed in an insect proof cage and pollination is effected through freshly reared houseflies.

c. **Marker gene:** The purple colour of new leaves and panicle and beak characters of fruit helps in identifying the hybrid seedlings in the nursery (Sharma and Majumder *et al.,* 1985).

d. A new off-season crossing technique was suggested by Kulkarni (1986). It involves induction of flowering in the desired parents in off season by veneer grafting, their defoliated shoots on to leafy shoots off season flowering *cv.* Royal Special and allowing open-pollination between the desired parents. As no other cultivar flowers during this season, this is a safe technique.

Promising Hybrids of Mango

- ☆ IARI, New Delhi: Mallika, Amrapalli, Pusa Arunima
- ☆ IIHR, Bangalore: Arka Anmol, Arka Puneet, Arka Aruna, Arka Nilkiran
- ☆ RFRS, Vengurla: Ratna, Sindhu, Konkan Ruchi
- ☆ CISH, Lucknow: Ambika
- ☆ FRS, Sangareddy: Au Rumani, Manjeera.
- ☆ HC and RI, Periyakulam: PKM-1, PKM-2

- ☆ BAC, Sabour: Safari, Jawahar
- ☆ AES, Paria: Neeleshan, Neeleswari, Neelphanso

In Israel, a new cultivar, Naomi, has been released which has smooth skin and red pigmentation. In Australia, a hybrid of Sensation x Kensington has shown promising results. In Israel, rootstock breeding is also in progress and a polyembryonic rootstock 13/1 has been released that is tolerant to salinity.

Mutation Breeding

Naturally occurring useful mutants like Rosica has been isolated from the Peruvian variety 'Rasado de lca'. Similarly, Davis Haden is a mutant of Haden. However, no induced mutant is known to have been released.

Polyploidy Breeding

Much scope exists for polyploidy breeding. However, till date there is no report on this line. Vellaikolumban cultivar of mango is tetraploid in nature (2n=4x=80) which is a polyembryonic type.

Heterosis

Iyer and Subramanyam (1984) observed large fruits in some progenies of Alphonso x Banganapalli. Transgressive segregation for this character was also observed. The population with bigger fruits was large among hybrid progenies obtained with Banganapalli as one of the parents. This effect may be due to an accumulation of dominant allele each having additive effects and masking the effect of deleterious recessive allele.

Biotechnology

Repetitive and proliferative somatic embryogenesis has been achieved using nucellar explants in monoembryonic varieties. High frequency somatic embryogenesis from nucellus tissue of monoembryonic mango Arka Anmol has been achieved. Media composition for conversion of embryogenic callus cells to embryos, their development, maturation and germination leading to the formation of complete plantlets has been worked out. Suspension cultures have been raised from nucellar callus in mango hybrid Amrapali to maximize production of somatic embryos.

Chapter 12

Banana

Botanical name : *Musa* sp.

Family : Musaceae

Chromosome number : 2n = 22, 33 or 44 also exists

History of Banana Breeding

Banana breeding was started in Trinidad, West Indies in 1922 and in Jamaica in 1924 (Shepherd, 1994). The driving force for this breeding programme was to develop improved *Fusarium* wilt (*Fusarium oxsyorum F.* sp. *cubense*) resistant banana for export trade. In 1960, both the programmes were combined under the Jamaica Banana Board. United Fruit Company also started a small breeding programme in Panama in 1920. In India hybridization work was started at Central Banana Research Station, Aduthurai, Tamil Nadu in 1949. Important banana growing states are Maharashtra, Karnataka, Kerala, Tamil Nadu, Andhra Pradesh, Odisha, Bihar, West Bengal and Assam. In recent days, in some districts of Uttar Pradesh, Harichal banana is cultivated on a commercial scale. In South India, other than its edible use, banana is extensively used in all auspicious occasions such as wedding, festivals and worshipping God. Banana is a good table fruit, besides, the cultivar Nendran is used for cooking. It is also used for preparation of chips.

Centre of Diversity

Edible banana is native to old world especially South East Asia. Malayan area seems to be the primary centre of origin of cultivated banana (*M. acuminata*). *M. acuminata* was probably introduced into India and Burma where *M. balbisiana* is a native species. Natural hybridization between these two species might have resulted in many hybrid progenies (AAB, ABB, *etc.*).

Botany

Banana is monocotyledonous plant. It consists of pseudostem, rhizome, leaf blades and inflorescence. Rhizome is the ideal stem on which large numbers of buds or eyes develop. Initiation of the inflorescence primordium of the banana bunch takes place in the heart of pseudostem.

Genus *Musa* has about 50 species and this genus is divided into five sections:

a) **Eumusa:** Includes about 13-15 species of edible and wild banana. The chromosome number is 2n=22 in wild species and most of the cultivated varieties are having 2n=33 (2n=44 rarely), *e.g., M. acuminata, M. balbisiana, M. basjoo, etc.*

b) **Rhodochlamys:** Mostly diploid, spread from India to Indonesia. Five to seven species are kept in this group. Parthenocarpy is absent in this group, *e.g., M. ornata, M. velutina.*

c) **Callimusa:** This is of ornamental value and x=10 and 2n =20. It is found in Indo- China, Malaya and Borneo. Parthenocarpy is absent in this type. It includes about 5-6 species, *e.g., M. coccinea.*

d) **Australimusa:** Like Callimusa it has x = 10 and 2n=20 chromosome. Species of this group is common in Queensland and Philippines. Important species of this group are *M. textilis* or manilahemp, *M. maclavi, etc.*

e) **Incertae sedis:** It includes *M. ingens* (x=7, 2n=14) of New Guinea which grows to a height of over 10 m. This is the largest known herb. Another species in this group is *M. beccarii* (x=9, 2n=18) from North Borneo.

Ensete is another genera of this family probably originated in Asia. Genus *Ensete* has 6-7 species of which *E. ventricosa* is reported to be grown in Ethiopia as a food crop. The most important *Musa* cultivars are almost sterile triploids (2n=3x=33) and also tetraploid and diploid banana cultivars have also local importance in Asia. All banana and plantain landraces are farmers selection from intra and inter-specific hybridization of two different species, *M. acuminata* Colta, donor of the A genome and *M. balbisiana* Colta, donor of the B genome. Simmonds and Shepherd (1955) reported scoring technique to indicate the relative contribution of the two wild species for the constitution of a given cultivar. Fifteen distinguishing characters between *M. acuminata* and *M. balbisiana* were identified by them. Score one was given for each character in which a cultivar agreed with *Musa acuminata* and scores five was given for each character to which agreed with *Musa balbisiana*. Intermediate expressions of the characters were assigned score of 2, 3, or 4 depending on their intensity.

Taxonomic Scoring of Banana Based on Distinguishing Features

Characters	*Musa acuminata*	*Musa balbisiana*
Pseudostem colour	More or less heavily marked with black or brown blotches	Blotches slight or absent
Petiolar canal	Margin erect or spreading with scarious wings below, not clasping pseudostem	Margins not winged below, clasping pseudostem
Peduncle	Usually downy or hairy	Glabrous
Pedicel	Short	Long
Ovules	Two regular rows in each locule	Four irregular rows in each locule
Bract shoulder ratio	Usually high (ratio:0.28)	Usually low (ratio:0.30)
Bract curling	Bracts roll	Bracts lift but do not roll
Bract shape	Lanceolate or narrowly ovate tapering sharply from the shoulder acute	Broadly ovate, not tapering sharply
Bract apex	Acute	Obtuse
Bract Colour	Red dull purple or yellow Inside pink, dull purple	Distinctive, brownish purple outside, bright crimson inside
Colour fading	Inside bract colour fades to yellow towards base	Inside bract colour continues to base
Bract scars	Prominent	Scarcely prominent
Free tepal of male flower	Variably corrugated below tip	Rarely corrugated
Male flower colour	Creamy white	Variably flushed with pink
Stigma colour	Orange or rich yellow	Cream, pale yellow or pale pink.

At the botanical garden, Howrah, seeds of few banana species were collected from Chittagong and Madras. More number of genotypes of banana was also maintained at Central Banana Research Station, Aduthurai. After that it was shifted to Horticulture College and Research Institute, Tamil Nadu Agricultural University, Coimbatore. After the formation of National Research Centre on Banana (NRCB), Trichy in 1995, a wide germplasm collection including wild types are being maintained at this centre and intensive research programmes are being taken up on various problems related with banana. Presently, Tamil Nadu Agricultural University is also maintaining 186 collections of germplasm.

Breeding Objectives

- To develop dwarf statured banana suitable for high density planting and to prevent damage from high wind velocity.
- Production of good quality fruits.
- Resistant to biotic and abiotic stresses, *i.e.,* nematodes, panama wilt, bunchy top, sigatoka leaf spot, moko disease and pseudostem weevil, *etc.*
- To develop varieties with wider agro-ecological adaptability.
- Development of male fertile parthenocarpic diploids with resistance to

major diseases and pests.

- ☆ Developing longer finger size.
- ☆ Suitability for export.
- ☆ Good keeping quality.

Taxonomic Classification of Edible Banana (Simmonds and Shepherd, 1955)

Genome	*Ploidy level*	*Score and nomenclature*
Constitution		
AA	2x	16-23 Matti, Anai Komban
AAA	3x	15-21 Gros Michel, Cavendish
AAAA	4x	15-20 Bodles Altafort (Synthetic hybrid of West Indies)
AB	2x	46-49 Ney Poovan, Kunnan
AAB	3x	26-46 Champa, Rsathali
ABB	3x	59-63 Kanchkela, Monthan
ABBB	4x	63-69 Klue Teparod

Breeding Methods and Achievements

Introduction

Introduction of some cultivars of banana was made with resistance to biotic stresses, *e.g.*, Lady Finger (EC 160160) resistant to bunchy top virus introduced from Australia and is being evaluated at IIHR, Bangalore and TNAU, Coimbatore. Further, cultivars Naine MS (EC 27237) from France and Valery from West Indies were introduced for utilization in improvement programme. Banana cultivar introduction called as G 9 (it also Grand Naine) in India has been found very successful commercially and Jain Irrigation System Ltd., Jalgaon is selling millions of tissue culture raised plants annually. The recent annual sale of such plants has been 8 crores plants, all raised through world class tissue culture and hardening facility at Jalgaon.

Hybridization

In India, breeding work was started at Central Banana Research Station, Aduthurai (Tamil Nadu) in 1949. Afterwards breeding programme was also initiated at TNAU, Coimbatore and Kerala Agricultural University, Trichur. Technique of hybridization in banana is different from other crops. Pollination is best carried out in the morning. The bunches of female parent are bagged at shooting and each successive hand is pollinated as it is exposed. At maturity and ripening the bunch is cut and seeds are extracted. Seeds are sown at once in the greenhouse.

Evaluation of hybrid progenies from seedlings to harvest may not be the correct phase instead, evaluation of the same under next vegetative phase, *i.e.,* sucker to harvest stage will be ideal as full expression of yield potential could be observed only in the second crop of the F_1 progeny. The first crop (seedling to harvest) takes more than 15-19 months, where most of the energy of the plants is needed for corm formation.

Three main approaches in breeding dessert bananas of the Cavendish types are:

1. 3n x 2n superior diploid; there is no chromosome reduction in the egg cells thus yielding tetraploids.
2. 4n bred tetraploids hybrids x 2n superior diploids producing 'Natural triploids'.
3. 2n meiotic restituting clones x 2n superior diploids producing 'Natural triploids'.

Evolution of Edible Banana

There are two wild species in banana, *i.e., Musa acuminata* which is denoted by AA and *Musa balbisiana* which is denoted by BB. They are cross naturaly and making zygote AB in banana some varieties are AB type. Due to result of polyploidy breeding make triploid and tetraploid variety. Tetraploid *acuminata* (AAAA) crossed with diploid *bulbisiana* and make AAB type (triploid) at the same time tetraploid *bulbisiana* cross with diploid *acuminata* and make ABB type (triploid).

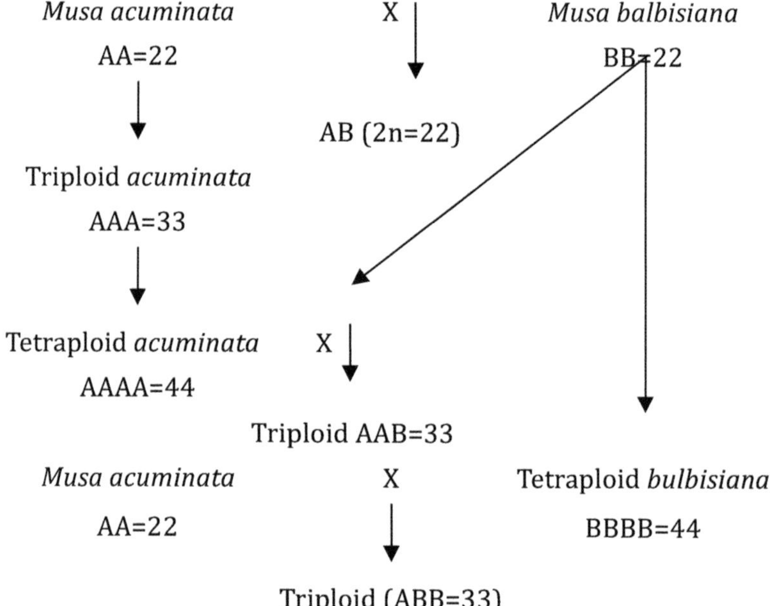

Developing New Diploid Male Parent

In many banana growing countries, initially wild diploid bananas (AA) were utilized as male parent and as a result, the resultant tetraploids had inherited many undesirable traits. Hence, it has been felt by banana breeders that the primary objective is to synthesize a good male parent. An ideal male parent must be highly resistant to Panama and Sigatoka diseases, must have vertical and compact bunch and fruits as large as the diploidy can allow and must be parthenocarpic having sufficient pollen to permit its use as a male parent. *Musa acuminata* subsp. *burmannica* and its hybrids offer a good source of resistance to Black Sigatoka. One such diploid developed in Honduras is SH 2989. Other male diploids worthy to be mentioned are SH 3142 for nematode resistance and SH 3176 evolved through multiple crosses for resistance to Black Sigatoka with desirable horticultural traits.

Breeding Work at TNAU

Since 1971, extensive inter-diploid crosses were made to synthesize new diploid forms at the Tamil Nadu Agricultural University, Coimbatore using the following parents: Matti (AA) is a diploid cultivar commercially grown in the southern most part of India. It exhibits a strong resistance to Sigatoka disease but is highly susceptible to nematodes. Its bunches weigh 12 to 19 kg with 9 to 10 hands containing fairly long fingers. It sets seeds when pollinated, though it is highly male sterile. This cultivar is extensively used as female parent in the diploid breeding programmes. *M. acuminata* subsp. *burmannica* has been shown to have resistance to fusarium wilt Races 1 and 2, Sigatoka diseases and nematodes.

Other diploid clones involved in the diploid male parent synthesis at Coimbatore are the indigenous cultivars Anaikomban (AA) and Namarai (AA). Anaikomban is resistant to nematodes and *Fusarium* wilt but susceptible to yellow sigatoka. It has long fingers (15-18 cm) and usually produces a smaller bunch weighing 6 to 8 kg. Namarai is a small slender plant, grown in Pulney and Sirumalai hills of Tamil Nadu with small fruits having piquant flavour and pleasant acid sweet taste. It has very short pedicel. It is susceptible to both Sigatoka disease and nematodes but no incidence of Panama disease is known so far.

The introduced diploids are Pisang Lilin (AA) and Tongat (AA), known for their resistance to panama disease and nematodes.

Many synthetic hybrids (diploids) have been developed which have good horticultural characters including resistance to Sigatoka, panama wilt and burrowing nematodes. These hybrids are now used as the male parents to cross with local triploid varieties or inter crossed to synthesize new triploid hybrids.

3n x 2n breeding programme taken up at TNAU has resulted in the development of CO-1 banana.

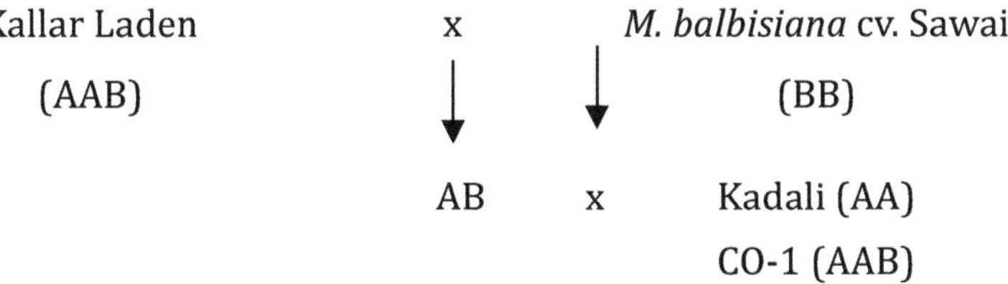

It is a pome group of banana of the genome AAB and closely resembles Virupakshi (AAB), a pome type banana popular in the hills of Tamil Nadu. At Kerala Agricultural University, two hybrids *viz.*, BRS-1 (Agniswar x Pisang Lilin) and BRS-2 (Vannan x Pisang Lillin) have been developed. BRS-1 (AAB) is 100 days earlier than Rasthali with significant differences in bunch weight. It has been released for homestead cultivation in Kerala, as it is resistant to Sigatoka leaf spot. BRS-2 (AAB) is a medium statured hybrid, tolerant to leaf spot and panama disease, rhizome, weevil and nematodes. The average bunch weight is 14 kg with 8 hands and 118 fruits in crop duration of 314 days.

Breeding Work in other Countries

PITA-9: A Black Sigatoka Resistant (BSR) hybrid from the "False Horn" plantain, a tetraploid hybrid having black sigatoka resistance has been developed at International Institute of Tropical Agriculture (IITA), Ibadan, Nigeria.'BITA-3' is a tetraploid starchy banana hybrid with low partial resistance to black Sigatoka disease developed at IITA High Rainfall Station in Onne (Southeastern Nigeria), where both (Banana streak virus) and cucumber mosaic virus (CMV) have been observed. 'BITA-3' is a hybrid from the interspecific cross 'Laknau' x 'Taju Lagada', 'Laknau' is a female fertile AAB starchy banana that closely resembles plantains. 'Taju Lagada' is an AA diploid banana having a long bunch with many hands. BITA-3' produces heavy bunches.

Banana breeding programme has been taken up in Honduras by the Fundacion Hondurena De Investigation Agricola (FHIA) with the aim of developing superior diploid plantations combining desirable agronomic traits with resistance, which is then used for production of primary tetraploids. This organization has developed many FHIA hybrids, which possess resistance to nematodes, *Fusarium* wilt, *etc.* Introduction and testing of these hybrids in India in various centers revealed superior performance of FHIA-1, FHIA-3, FHIA-21 and FHIA-25.

Mutation Breeding

Bud mutation in Indian banana is very common perhaps due to spontaneous rearrangement of chromosomes in somatic meristem and structural re-assortment. A great majority of edible bananas are triploids, a condition that interferes with normal equilibrium of plants and may provide the requisite stimulus to structural rearrangement of chromosomes, leading ultimately to the evolution of a new gene complex. Several natural sports of well established commercial clones have been recognized, *e.g.*, High Gate (AAA) is a semi-dwarf mutant of Gros Michel (AAA), Motta Poovan (AAB) is a sport of Poovan (AAB), Ayiranka Rasthali is a sport of Rasthali (or Silk), Barhari Malbhog is a sport of Malbhog, Krishna Vazhai is a natural mutant of Virupakshi (or Pome), and Sambrani Monthan (ABB) is a mutant of Monthan (ABB).

In Nendran, more than six mutants have been recognized. One of these, Moongil, has undergone such a radical change that there is no male phase and a bunch has only one or two hands with biggest size fruits. Attu Nendran, Nana Nendran,

Myndoli, Velathan and Nenu Nendran are a few mutants which have been selected for one or the other desirable character. Similarly, Ambalakadali and Erachi vazhai are mutants of Red Banana. The Kunnan variety of Malabar has provided a few mutants known as Thattilla Kunnan (male phase absent), Veneetu Kunnan.

Adakka Kunnan and Thaen Kunnan from *cv.* Monthan, Sambal Monthan, Nalla Bontha Batheesa, Sambrani Monthan, Pidi Monthan and Thellatti Bontha have been recognized as sports. At INIVIT, Cuba induced mutations of ABB cooking banana, Burrow Cemsa, was obtained. At TBRI, Taiwan, Tai Chiao 1 and GCTV, triploid bananas with *Fusarium* wilt resistance are obtained as a result of clonal variation of AAA Cavendish banana. The early flowering FATOM 1 was developed as a result of *in-vitro* gamma irradiated meristem culture of *cv.* Grand Naine has been released in Malaysia.

Biotechnology

Plant tissue culture and molecular biology techniques are applied to enhance the handling and improvement of banana. Important application of a cell biology are micro-propagation for rapid multiplication and germplasm exchange, embryo culture/rescue for *in-vitro* seed germination, cryopreservation of germplasm and genome manipulation through genetic engineering using cell suspensions or protoplast culture. Although, somaclonal variation through micro-propagation is of limited use in plantain breeding, it has been successfully applied in Taiwan for the development of improved Cavendish banana cultivars with resistance to *Fusarium* wilt and acceptable fruit quality. In gene transfer methods, Sagi *et al.* (1995), from Katholieke University, Leuven, Belgium reported the transgenic triploid cooking banana showing transient expression of GUS marker gene in pot growing in the green house from DNA particle bombardment on ABB cooking banana. The molecular markers are providing tools for phylogenetic investigations and cultivar identification, basic genetic research, marker assisted selection and diagnostics in pathogen identification.

Chapter 13

Papaya

Botanical name : *Carica papaya* L.

Family : Caricaceae

Chromosome number : 2n = 2x = 18

History of Papaya Breeding

Papaya is an ideal fruit crop for growing in kitchen garden, backyards of home as well as orchards, especially those places nearer to cities or big town. It is also grown as a filler plant in orchard. The ripe fruits of papaya are consumed throughout the tropics and subtropics. Fruits are also used in preparation of jam, soft drinks, ice-cream flavour, crystallized fruits and syrups. Unripe fruits are commonly used as vegetables. Papain is prepared from the latex of immature fruits. It is a proteolytic enzyme used for tenderizing the meat as well as in leather tanning and cosmetics, *etc.* The ripe fresh fruit is a rich source of vitamin A (2020 IU), vitamin C (40-60mg/100g), carbohydrates and minerals. Papaya, on account of producing fruit in a short period after planting has attracted the attention of fruit growers for large-scale cultivation in the country. It is known as wholesome fruit which is valued for its nutritional and medicinal properties. Due to its nutritional, industrial and export demand, papaya is recognized the most potential fruit crop for commercial cultivation. The variable sex forms, susceptibility to frost and water-logging, fungal and viral diseases are well identified problems in its cultivation. Economically the productive life span of papaya is 3 years and every time new plantation has to be raised.

Centre of Diversity

Papaya is native to Tropical America. The South America and Costa Rica is the micro centre of origin of papaya. It is a close relative of *Carica peltata*. In India, it was introduced in the early part of the 16th century from Philippines through Malaysia. It was widely spread in different parts of the country particularly tropical and sub-tropical zones. India is the largest producer of papaya in the world. It is also cultivated in Brazil, Mexico, Australia, Hawaii, Malaysia, Taiwan, Peru, Florida, Gold Coast, South Africa and Bangladesh. In India it is widely cultivated in Uttar Pradesh, Bihar, Karnataka, Jharkhand and Madhya Pradesh.

Botany

Papaya belongs to family *Caricaceae* and genus *Carica* having about 40 species. It is a dioecious plant but gynodioecious cultivars are also available. The stem is hollow and soft wooded. It is usually unbranched when young but at later stage upright shoots develops at its terminal growth due to obstruction. The leaves are palm like with long stalks. Flowers are cymose, fragrant borne in leaf axils. The fruit is a fleshy berry. The fruits from pistillate flowers are ovoid-oblong to nearly spherical in shape and the fruits from hermaphrodite flowers are pyriform, cylindrical or grooved.

Floral Biology and Pollination

Dioecious papaya produces male and female trees separately on different plants in the ratio of 1:1, while gynodioecious cultivars produces both female and andromonoecious trees in the ratio of 1:2. Female and male flowers develop within 32 and 42 days respectively after bud initiation. The period from bud initiation to anthesis is shorter for male than female flower bud. Stamen development occurs prior to ovary development in the hermaphrodite flower and stamen differentiation is observed 56-59 days before anthesis. Anther dehiscence is completed within 18-36 hours before the flowers opening and continues depending upon the weather conditions and stigma becomes receptive a day before the flower opening and remaining receptive for 6 days. The peak anthesis was observed between 5.00-6.00 am. The receptivity of stigma was found maximum on the day of anthesis in most of the species. To ensure good fruit set in the dioecious cultivars of papaya (*e.g.*, CO-1, CO-2, CO-4, CO-5, CO-6, Pusa Giant, Pusa Dwarf, Pusa Nanha), the female and male ratio should be 20:1. However, maximum numbers of andromonoecious trees are retained in gynodioecious varieties (*e.g.*, CO-3, CO-7, Coorg Honey Dew, Sun Rise Solo, Sun Set Solo, *etc.*). Further, insects are the major pollinating agents in papaya.

Sex Forms

There are two major sex forms in papaya:

1. **Dioecious:** Male and female trees segregate in the ratio of 1:1, sibmating is done for maintaining of purity.
2. **Gynodioecious:** Female and andromonoecious (male+bisexual flowers in a single tree) trees segregated in the ratio of 1:2. Selfing of bisexual flowers is done for obtaining pure seeds.

Germplasm Resources

The family Caricaceae consists of six genera and 35 species. *Carica* and *Vasconcellea* are the important generas. The genus Carica has only one species, *Carica papaya* the cultivated species. *Vasconcellea* contain 21 species, which are considered as the wild relatives of papaya. Their diversity is common in South America. Variations with respect to plant stature, sex types, fruit shape and size, seed content are observed in papaya. Presently, germplasm is being maintained at TNAU, Coimbatore, IIHR, Bangalore, IARI Regional Station, Pusa, Bihar, CHES, Ranchi, CHES, Bhubaneshwar and CISH, Lucknow for further characterization and evaluation.

Important Wild Relatives of Papaya

Some *Vasconcellea* species are used in interspecific hybridization for resistance breeding, *e.g., Vasconcellea Carica cauliflora* is resistant to viruses. *Vasconcellea candamarcensis* and *Vasconcellea pentagona* are resistant to frost. Based on the crossability and compatibility it has been observed that *Vasconcellea monoica, Vasconcellea cauliflora* and *Vasconcellea candamarcensis* are easily crossable with each other and producing viable seeds. But inter-generic hybridization of these species with *Carica papaya* did not produce mature seeds. However, by using embryo culture technique, immature embryo can be developed into mature embryo. Further, cross between *Carica papaya* and *Vasconcellea goudotiana* was found to be a failure. The species, *Vasconcellea cauliflora, Vasconcellea pubescens, Vasconcellea stipulata* and *Vasconcellea candicans* are resistant to papaya ringspot virus.

Breeding Objectives

- ☆ To develop dwarf statured and early bearing varieties.
- ☆ To evolve varieties with high yield and good quality fruits.
- ☆ To develop varieties with low cavity index and more pulp thickness.
- ☆ To breed varieties having good keeping quality and suitable for export.
- ☆ Breeding for high latex yield with high proteolytic enzyme activity.
- ☆ To develop varieties resistant to biotic and abiotic stresses (virus, frost, water logging, *etc.*).

Breeding Methods and Achievements

Introduction

Commercial cultivars such as PI 41436, Soniyimma, Malinchly, Maru Ank LG (Nigeria), Sunrise, Wilder, Sunset, Kapohosolo, and California (USA) have been introduced. The cultivar California is dioecius in nature with high degree of tolerance to papaya ringspot virus.

Inbreeding and Selection

In dioecious lines, suitable male plants are selected from the same progeny which have resemblance to female plants in vegetative characters, such as stem and leaf colour, stem thickness and height at flowering, *etc.* Progenies raised from S_1 inbreds are screened and desired male and female plants are selected for further sibmating, *i.e.,* crossing between the female plant and male plant of the same cultivar. The process is to be continued for 7-8 generations to achieve uniformity for a group of characters. In this method, the progeny will have male and female in equal proportion. Many dioecious cultivars have been bred by this method.

Development of cultivar with high papain content was started at TNAU, Coimbatore. As a result of intensive breeding programme, four cultivars, *i.e.,* CO-1, CO-2, CO-5 and CO-6 were developed through inbred selections which are dioeicous. Recently during 2011, CO-8 papaya was developed in the dioecious group through sib-mating of CO-2 variety, which is a red-fleshed variety and this variety is unique for the red pulp colour which does not exist in any of the dioecious papaya varieties developed. Systematic work for breeding of uniform varieties with high yield and good quality for wider adaptability was started at IARI Regional Station, Pusa, Bihar in 1966. As a result of inbreeding and selection for 8 generations during 1966-1982, uniform lines of Pusa Delicious, Pusa Majesty, Pusa Giant and Pusa Dwarf with desirable attributes were developed. At Coorg, Aiyappa and Nanjappa (1959) selected Coorg Honey Dew which is gynodioecious. Based on two seasons evaluation of papaya germplasm in Nainital, cultivars, namely, Barwani Red, Coorg Honey Dew, BR 1, 2, 3, 4, Coimbatore 1B and Washington have been found promising for commercial cultivation. Considerable genetic variability was found in many plants with respect to fruit characteristics like fruiting height, fruit weight, size shape, TSS, taste and number of fruits per tree.

Important Cultivars of Papaya Developed through Selection are as under

Varieties	*Characters*
CO-1	Plant is dwarf; fruit is round, selected from cultivar Ranchi.
CO-2	It is pure line selection from local type, suitable for papain production in terms of high enzyme activity. Dual purpose variety for dessert fruit as well as papain.
CO-3	It is a hybrid between CO-2 x Sun Rise Solo, plant is vigorous,fruit is medium in size with good keeping quality.
CO-4	It is a hybrid between CO-1 x Washington, fruit is large, keeping quality is good, and flesh colour is yellow.
CO-5	It is a selection from Washington; it is also good for papain production in terms of high latex yield.
CO-6	It is a selection from 'Giant. Fruits are bigger weighing 2.5 to 3.0 kg. Suitable for papain and dessert purpose.
CO-7	It is gynodioecious hybrid between CP75 x Coorg Honey Dew. It has red flesh. Fruit cavity round.
CO-8	It is a red-fleshed dioecious variety bred through selective sib-mating in the population of CO-2. Suitable for dessert purpose, processing and papain production.

Varieties	Characters
Coorg Honey Dew	It is gynodioecious selection from Honey Dew at Coorg.
Pusa Majesty	It is selection from *cv.* Ranchi. It is gynodioecious variety suitable for high papain.
Pusa Delicious	A gynodioecious selection from *cv.* Ranchi, it has good fruit quality.
Pusa Giant	A dioecious selection from *cv.* Ranchi, it is tolerant to strong wind.
Pusa Dwarf	It is a selection from *cv.* Ranchi, plant is dwarf and dioecious in nature, fruit shape is oval.
Pusa Nanha	It is developed through mutation; plant is dwarf, suitable for high density planting, fruit quality is good.
Surya	It is a gynodioecious hybrid (Sun Rise Solo x Pink Flesh Sweet), developed at IIHR, Bangalore, it has high yield with good quality fruit.

In recent times, an introduction from Taiwan via Thailand by private seed companies sold as Red Lady is becoming very popular among the farmers and the consumers. Most of the public sector varieties as listed above have gone out of cultivation because of non-availability of seed.

Induction of Polyploidy

Hofmeyer (1945) reported on polyploidy in papaya. They found that the quality of tetraploid fruit was better than the diploid and it was also compact with small seed cavity. But tetraploids were less fertile than diploid as indicated by comparative seed count. However, according to Singh (1955) there was complete sterility in both female and male tetraploids and expressed doubt about their commercial utilization. Further, Zerpa (1957) reported that colchicine induced tetraploid hermaphrodite plants, which were used as male parent in a cross with a female diploid produced a few seeds without endosperm, by embryo culture, two triploid plants were obtained which turned out to be hermaphrodite.

Hybridization

A few hybrid varieties have been developed by the inter-varietal or inter-generic hybridization. But still there is great scope for development of superior cultivars with better quality and yield. At TNAU, Coimbatore three varieties have been developed, *viz.,* CO-3 (CO-2 x Sunrise Solo), CO-4 (CO-1 x Washington) and CO-7 (CP.75 x Coorg Honey Dew). At IIHR Bangalore, two hybrids IIHR-39 named as Surya (Sun Rise Solo x Pink Flesh Sweet), and IIHR-54 (Waimanalo x Pink Flesh Sweet) were developed (Dinesh and Yadav 1998), Hybrid HPSC-3 (Tripura Local x Honey Dew) was developed by the ICAR Research Complex Tripura (Singh and Sharma, 1996). Cultivar Cariflora was developed by crossing K_2 x K_3 line of papaya which is tolerant to PRSV. However, seed availability is a big issue for these cultivars.

Inter-specific hybridization was also attempted in the genus *Carica*. The cross between the *Carica papaya* and *Carica caulifora* did not form mature seed but immature embryos could be germinated and grown by embryo culture. F_1 hybrids of *Carica papaya* x *Carica pubescens* and *Carica papaya* x *Carica quercifolia* were vegetatively vigorous. *Carica papaya* x *Carica cauliflora* F_1 progenies are slow growing and those of *Carica papaya* x *Carica stipulata* developed apical necrosis before reaching maturity. Iyer and Subramanyam (1984) attempted interspecific hybridization and reported that F_1 hybrids of *Carica cauliflora* x *Carica monoica* when crossed with *Carica papaya* gave fertile hybrids, although the *Carica cauliflora* and *Carica monoica* are incompatible with *Carica papaya* but hybrids of *Carica cauliflora* x *Carica monoica* are compatible.

Heterosis Breeding

Dai (1960) reported heterosis in the cross between Philippines x Solo varieties. F_1 hybrid tended to have reduced seed number and enhanced plant vigour. Heterosis up to 111.4 per cent for yield and yield traits was obtained in Solo Yellow x Washington whereas high heterosis for potential economic competitiveness was noticed in Thailand x Washington (Iyer and Subramanyam, 1981). At IIHR, Bangalore, an F_1 hybrid namely, Surya (Sun Rise Solo x Pink Flesh Sweet) was released recently. It is gynodioecious in nature and produces about 75-80 fruits of medium size weighing about 600-800 g per fruit. The flesh is red in colour, firm and sweet to taste with a TSS of 14 °Brix.

Mutation Breeding

Ram and Majumder (1981) developed a dwarf mutant line by treating papaya seed with 15k gamma rays. Initially, 3 dwarf plants were isolated from M_2 population. Repeated sibmating among the dwarf plants helped in establishing a homozygous dwarf line Pusa Nanha. However, it could not make headway due to non-availability of seed after retirement of original breeder, Dr. Mansha Ram.

Biotechnology

In-vitro propagation and genetic engineering technique can serve as a potential tool to overcome major constraints in *Carica papaya.*

Embryo Culture

In papaya, incompatibility is mainly due to the failure of endosperm formation. The hybrid embryo resulting from interspecific cross of *C. papaya* and *C. cauliflora* has been successfully rescued on white's medium in 30 days.

Transgenic Papaya

Transgenic papaya has been developed against Papaya Ring Spot Virus (PRSV) using coat protein mediated resistance in University of Hawaii by Dennis Gonsalves. The coat protein gene from PRSV was isolated, cloned and used for transforming papaya to provide resistance against the severe strain of the same virus. The target cultivars used in transforming papaya were the Red Fleshed, Sun Set Solo and the Yellow Fleshed Kapoho Solo. Transformation with coat protein gene was done using micro projectile bombardment technique using embryogenic tissues of papaya.

Two transgenic lines Sun UP from Sunset and UH Rainbow from Kapoho were developed which have shown excellent resistance to PRSV. Sun UP, which is homozygous for CP (Coat Protein gene), was resistant to most isolate of PRSV, from other geographical locations except Taiwan's YK isolate of PRSV. Rainbow was found susceptible to PRSV isolates from outside Hawaii but was resistant to the severe strain of Hawaiian PRSV (HA isolates).

Improved Varieties

Coorg Honey Dew, CO-1, 2,3,4,5,6, Pusa Majestic, Pusa Delicious, Pusa Dwarf, Pusa Nanha, Surya, Punjab Sweet, *etc.*

Chapter 14

Guava

Botanical name : *Psidium guajava* L.

Family : Myrtaceae

Chromosome number : 2n=2x=22

History of Guava Breeding

Guava is also known as the 'Apple of the Tropics'. It is a very rich and cheap source of vitamin C and also contains a fair amount of calcium. Important guava growing states in the country are Uttar Pradesh, Bihar, Madhya Pradesh and Maharashtra. Allahabad district of Uttar Pradesh has the reputation of growing the best quality of guava fruits in the world. The importance of guava is due to the fact that it is the hardy fruit which can be grown in alkaline and poorly drained soil.

Centre of Diversity

Tropical America is supposed to be the centre of origin of guava where it is found in wild as well as cultivated forms. Guava came to India at a very early time before 17th century.

Botany

Most of the cultivars of Indian guava belong to the genus *Psidium* and species *guajava.* Based on the shape of common guava fruits, they are classified into two groups (De Candolle 1904), *i.e., Psidium pyriferum*, *Psidium pomiferum*. Genus *Psidium* contains about 150 species (Hayes, 1970). All cultivated varieties of guava are either diploid 2n-2x-22 or triploid 2n-3x-33.

Floral Biology and Pollination

Guava bears flower solitary or in cyme of two to three flowers, on the current season growth in the axil of the leaves. About one month is required from flower bud differentiation to complete development upto calyx cracking stage. Peak time of anthesis is between 5.00-6.30 am in most of the varieties of guava. The dehiscence of anthers starts 15-30 minutes after anthesis and continues for two hours. The pollen fertility is high in almost all the cultivars. The pollen fertility is 78 per cent and 91 per cent in Allahabad Round and Lucknow Safeda, respectiviely.

Inheritance Pattern

- Bold seed is found to be dominant over soft seed and governed monogenically.
- Red flesh colour is dominant to white pulp colour and also goverened monogenically.
- Red fleshed cultivars are supposed to be heterozygous.
- There seems to be linkage between red flesh colour and bold seed size.
- Triploidy and some other genetic factors are responsible for female sterility.

Germplasm Resources

Guava is mainly a self-pollinated crop but occurrence of cross-pollination results in great variation in the seedling population. About 103 genotypes are available in the Indian collections has listed 153 genotypes including *Psidium* species, cultivars and hybrids mainly at CISH, Lucknow, IIHR, Bangalore, NDUA & T, Ayodhya, and HAU, Hisar. Guava germplasm is being maintained at several centers in the country in field banks which are often not systematically maintained. GBPUA & T, Pantnagar has good collection of guava germplasm.

Breeding Objectives

- Devlopment of less seeded variety.
- Less pectin content for edible purpose.
- More pectin content for processing.
- Uniform ripening.
- High keeping quality.
- Resistance to tea mosquito bug and wilt.

Breeding Methods and Achievements

Introduction

There are number of varieties and species introduced from different countries from time to time for species purpose. For examples, Beaumont and Indonesian Seedless from Australia, Acerapera from Brazil, Verdie from USA, *etc.*

Clonal Selection

Propagation by seeds during early days gave rise to considerable variation in the form and size of fruit, the nature and flavour of pulp, seediness and other morphological characters such as spreading or erect growth habit of the tree. Improvement work in guava was started for the first time in the country in 1907 at Ganesh Khand Fruit Research Station, Pune primarily with the collection of seeds of varieties, grown in different places to isolate superior strains. About 600 seedlings were raised and evaluated for fruit and yield characters. One strain from open-pollinated seedlings of Allahabad Safeda collected from Lucknow was selected and released as Lucknow-49 which is a popular variety throughout India.

At Horticultural Research Station, Saharanpur, evaluation of seedling types resulted in a superior selection, S-1, having good fruit shape, few seeds, sweet taste and high yield.

At Narendara Deva University of Agriculture and Technology, Ayodhya, out of the 23 strains collected as a result of survey in guava growing region, 3 seedlings of Allahabad Safeda (AS1, AS2, AS) and 2 of Faizabad Selections (FS1 and FS2) were found to be promising with respect to fruit quality and yield. At IIHR, Bangalore, from 200 open-pollinated seedlings of variety, Allahabad Safeda collected from Uttar Pradesh, one seedling selection, selection-8, was found to be promising. These plants are dwarf and give higher yield. The fruits are of medium size with white pulp and few soft seeds and excellent taste. This selection has been named as Arka Mridula.

Achievements

Varieties	*Important Characters*
L-49	Developed at GFES, Pune, and Seedling selection of Allahabad Safeda, semi-dwarf tree, high yielding and white flesh.
Banarsi Surkha	It is a selection from local red fleshed type, heavy bearer, large fruits, flesh soft and pink.
CISHG-1	Developed at CISH, Lucknow. Fruit skin colour is deep red, TSS 15°Brix, soft seeds.
Bangalore Local	It is a local selection, with white flesh and soft seeds, fruit are large.
Arka Mridula (Sel -8)	Developed at CISH, Lucknow, it is a selection from apple colour seedling, skin and flesh colour is pink with good acid sugar blend.
Plant Prabhat	Seedling selections from GBPUA and T, Pantnagar, prolific bearer, soft seed with good quality, flat-round white colour, smooth skin fruits.
Lalit (CISHG-3)	Developed at CISH, Lucknow, it is selection from apple colour seedlings, skin and flesh colour is pink with good acid sugar blend.

Hybridization

At IIHR, Bangalore, as a result of hybridization among Allahabad Safeda, Red Flesh Chittidar, Apple Colour, Lucknow-49 and Bananas, 600 F_1 hybrids were raised.

One hybrid Arka Amulya has been released recently. It is a progeny from the cross Allahabad Safeda x Triploid. Plants are medium in vigour and are spreading type. Fruits are round in shape. Skin is smooth and yellow in colour. Fruits on an average weigh about 180-200 g, Flesh is white in colour and firm. TSS is around 12° Brix, soft seeded, keeping quality is good.

Hybrid 16-1 (Apple Colour x Allahabad Safeda) has been developed. Plants are semi vigorous, moderate yielding, fruit skin bright red with few seeds high TSS and good keeping quality. At Fruit Research Station, Sangareddy (Andhra Pradesh), inter-varietal hybridization resulted in the isolation of two superior hybrids.

Safed Jam: This is a hybrid between Allahabad Safeda and Kohir (a local collection from Hyderabad–Karnataka region). It is similar to Allahabad Safeda in growth habit and fruit quality. The fruits are bigger in size with good quality and few soft seeds.

Kohir Safeda: It is a hybrid between Kohir x Allahabad Safeda. Tree is vigorous, fruits are larger with few soft seeds and white flesh. Haryana Agricultural University, Hissar has released two hybrid varieties.

Hisar Safeda: It is a cross between Allahabad Safeda x Seedless, which has upright growth with a compact crown. Its fruits are round, weighing about 92 g each, pulp is creamy white with less seeds, which are soft, TSS is 13.4 per cent and ascorbic acid 185 mg/100g.

Hisar Surkha: It is a cross between Apple Colour x Banarasi Surka. Tree is medium in height with broad to compact crown fruit is round weighing 86 g each. Pulp is pink having 13.6 per cent TSS, 0.48 per cent acidity and 169 mg/100g ascorbic acid. Yield is 94 kg/tree/year. CISH, Lucknow isolated two hybrids H-136 for red pulp and soft seeler with high TSS.

Few private nurseries in Malihabad, near Lucknow are multiplying and supplying plants of a local variety called as Baraf Khan, which is certainly a seedling selection but looks promising and has acceptance in the local farmers.

Breeding for Wilt Resistance

Work at CISH, Lucknow has shown that Chittidar, Portugal, Seedless and spear acid are tolerant to wilt. Resistance species of guava can be utilized for imparting the wilt resistant character. It was observed that *Psidium guajava* and *Psidium chinensis* are compatible. However, cross between *Psidium guajava* and *Psidium molle* was incompatible but reciprocal combination was a compatible combination.

Polyploidy Breeding

Producing triploids will be futile since the fruit shape in triploid is highly irregular and misshapen because of differential seed size. However, in order to evolve varieties with less seeds and increased productivity, crosses were made at IARI, New Delhi, between seedless triploid and seeded diploid variety Allahabad

Safeda of the 73 F_1 hybrids raised 26 were diploids, 9 trisomics 5 double trisomics and 13 tetrasomics. Distinct variation in tree growth habit and leaf and fruit characters was observed. Three trisomic plants had dwarf growth habit and normal shape and size of fruits with few seeds. The imbalance in chromosome numbers in aneuploids imparted sterility resulting in seed reduction in fruits.

Characteristics of Important Species and Cultivars

Psidium guineense: This is also known as the Guinea guava or Brazilian guava. The plants are like shrub or small tree. The leaves are green in colour, broad, oblong-oval, acute or obtuse, 8-12 cm long with lower surface pubescent. Red hairs are found on the mid veins.

***Psidium montanum*:** Plants are just like shrub, attain a height of about 1.5m. It is found in mountains of Jamaica. Fruits are round with very poor quality

Psidium friedrichsthalianum: It is known as Chinese guava. Plants are tall (7-11 m height), fruits are small and globose in shape with high acid content. It can be used for jelly making. Plants are tolerant to guava wilt.

Psidium cattleianum: It is known as the Cattely guava or Strawberry guava. It is a shrub or small tree (3-6 m in height), fruits are small, deep scarlet in colour, and globose in shape. This species is more tolerant to low temperature than *Psidium guajava.*

Psidium cattleianum* var. *lucidum: Tree height is more than Cattley guava in Hawai Island. Height of plant is noticed up to 12 m. Generally, it is propagated through seeds. Fruits are yellow in colour and used for jelly making.

Psidium molle: Tree is medium in height; leaves are green and oval in shape. Apex of leaf is pointed; lower part of leaves is velvety in appearance. Red hairs are found on the central veins. In one leaf 6-8 pairs of primary veins are found. Petals are 5-11, stamens are 196- 239, and stigma is long with big ovary of 3-5 chambers. Fruits are small in size, average fruit weight is 13 g. It contains vitamin C about 70 mg/100 g of pulp.

Psidium pumilum: It is also known as Chinese guava. Tree is like pyramidal in shape, leaves are light in colour, small in size, non-pubescent, having 13-17 pairs of primary veins. Petals 7 smooth and creamy colour which drop immediately after anthesis, Stamens are 252-327 in number, small stigma with medium size of ovary having 4-5 chambers. It flowers twice in a year. It takes about 130 days for attaining the maturity of fruits. Average fruit weight is about 19 g and average vitamin C content is 171mg/100 g of pulp.

***Psidium cujavilis*:** Growth characters and flowering habit of the plants is just like *Psidium guineense*. The size of fruit is small to medium, average weight is 30-50 g, and sour in taste.

Psidium policarpum: The growth characters are similar to *Psidium guajava* except the shape of the fruits is periform. Average fruit weight is about 200-250 g. Flavonoid patterns show close affinity between *P. guajava and P. molle*. However, in spite of the morphological similarities in *P. molle* and *P. guineense,* they showed minute differences in flavonoid pattern.

Chapter 15

Citrus

Botanical name: *Citrus* sp.

Family: Rutaceae

Chromosome number: 2n = 2x = 18

History of Citrus Breeding

Citrus constitutes a major group of fruits comprising of mandarins, oranges, lemon, pummelo, grape fruit, tangelo, trifoliate orange, citron, citranges, *etc.* Despite inter-specific and inter-generic hybrids, *Poncirus* and *Fortunella* also belong to genus *Citrus*. During its long history, citrus has given the world numerous varieties both by open pollination, bud sports and of recently by controlled pollination and artificial induction of bud variation. Citrus fruit cultivation lies between latitude 40°N also 40°S where conditions are neither cold nor moist and dry. India is considered to be the home of several citrus species and they are found growing wild in some parts of the country. Many types of citrus still remain unexploited by man and such types are considered as semi-wild.

Centre of Diversity

There are three major centres of diversity in India. The first is in the North-East including Assam and adjoining areas. It includes papedas, pummelos and their hybrids, citron, lemons and mandarins and other interesting types like jenera-tenga, soh synteng, a sour fruit similar to the sweet lime and soh siem, a mandarin type. The second diversity in south India, indigenous types include Gajanima, kichili and some wild mandarin types. The third is in North-West region at the foot of Himalayas where the hill lemon (galgal) is common. The various types of mandarins, hybrids of pummelo, citron, lemons, karna-khatta and rough lemon are found all over the country. In general, the wild types are more common in the foot hills. Many of the progenators of citrus fruits are believed to have originated in India. These include *C. latipes, C. limonia, C. kama, C. pennevesiculata, C. maderaspatana,* many of these are wild types. Presence of Sah-Niangriang, a wild sweet orange and a wild mandarin (*C. indica*) furnishes strong evidence that Eastern India might be the centre of orgin for many citrus fruits.

Floral Biology

Flowering in citrus takes place during February–April. In North India, sweet orange and mandarins bloom only once in March. However, it is reported that sweet oranges bloom twice in a year under Bihar conditions, *i.e.,* February–March and June–July. Inflorescence in citrus species is of cymose type. Generally anthesis takes place in the morning between 9.00 am to 12.00 noon. Flowers on shaded side of the tree have been observed to open later than those exposed to sunshine.

Germplasm Resources

Exotic collection of citrus germplasm was started in 1940. Kinnow mandarin was one of the collections which is now a leading cultivar in North–Western India. Besides, other exotic collections were Valencia Late, Washington Navel, Jaffa, Malta Blood Red, Pineapple, Ruby orange, Satsuma, Dancy Tangerine, Climentime, and Cleoptera wilking, Temple, Duncan, Marsh seedless, Lisbon lemon, Trifoliate orange, Dancy, Lisbon lemon, Trifoliate orange, (Dutta,1958), More than 650 accessions are being maintained at CHES, Chethali, Bangalore, CHES, Ranchi, RFRS, Abhor, NRC on citrus, Nagpur, Horticultural Experiment Station, Bathinda, IARI, New Delhi, MPKV, Rahuri, Citrus Improvement Project, Tirupati, Citrus National Research Centre, Nagpur, HC and RI, Periyakulam, and Citrus Experiment Station, Tinsukia, Assam. During 1988 as a result of systematic exploration by NBPGR in North-Eastern region, *C. indica* and many endangered species were collected for conservation.

North-Eastern region is a hunting ground of biodiversity of *Citrus* species. Chakrawar *et al.* (1988) identified two promising clones of acid lime Vikram and Pramalini in Maharashtra. At Nagpur, seedless Santra has been selected which gives high yield and quality fruits.

Attempt has been made during 1978 by NBPGR to preserve the *C. indica* which is progenitor of *C. reticulata*. For establishment of gene sanctuary, National Park, the natural genetic diversity of *C. indica* was observed in the forest of Garo hills in Megalaya which exhibited plant characters varying from bush to climber with high frequency of distribution in dense forest and showing resistance to biotic stresses. Therefore, a gene sanctuary for *C. indica* was established in Tanga Range in Garo hills. Genetic material of citrus is conserved in field gene bank or repository.

Problems in Citrus Breeding

There are three major problems which hinder the success of citrus breeding.

Time

Citrus being perennial in nature takes more time for bearing. However this period can be reduced to a maximum of half by top working the seedling on an old tree.

Polyembryony

It is peculiar feature found in citrus in which seed consist of more than one embryo. In addition to the zygotic embryo, one or more sometimes as many as fifteen additional embryos are developed from the nucellar tissue called nucellar embryos and found in the embryo sac. Most often, the zygotic seedling is crowded out by the vigorous nucellar seedlings. Forgetting large number of hybrids, citrus breeder should select a seed parent known to be either monoembryonic citrus species or polyembryonic except (*C. medica, C. latifolia* and *C. grandis*) which are monoembryonic restricts the choice of breeder and complicate the procedure required to attain the desirable objectives.

Sterility

Sterility is inability of gametic or sexual reproduction. Prevalence of high generative sterility is obviously a serious hindrance in the use of a particular parent for hybridization. Complete pollen sterility is problematic, where proportions of nucellar embryos are very high. High level of sterility often leads to production of seedless fruits which is serious hindrance to develop varieties.

Self-Incompatibility

Self-incompatibility and cross incompatibility is a common phenomenon which occurs widely in citrus. Most of the varieties of grape fruit (*C. grandis*) are found to be self incompatible besides, some varieties of lemon, sweet orange and mandarins exhibit self-incompatibility of gametophytic type goverened by oppositional alleles. Hybrid cultivars including Clementine, Orlando, Minneola, Sukega, Nova, Robinson are cross incompatible. Nova and Robinson is also suspected to be cross incompatible. Sweet orange varieties like Washington Navel and Satsuma mandarin are having sterile pollen; thereby they produce parthenocarpic fruits if cross-pollination is not done through viable pollens.

Long Juvenility

It is a major barrier in the progress of citrus breeding in India. General treatments to shorten the period or induce early flowering have not been generally effective. It was reported that neither chemical treatments nor incorporation by genetic transfers has been effective in combating long juvenility in citrus.

Breeding Objectives

- ☆ Producing early maturing citrus fruits with high yield and fruit quality.
- ☆ Developing rootstocks having disease and nematodes resistance, wider adaptability, *etc.*

In rootstock breeding, the main emphasis has been given on the development of root stock resistant to tristeza virus, *Phytophthora*, nematodes, *etc.* Most of the breeding programmes make use of *Poncirus*, which is a carrier of resistance to tristeza, *Phytophthora* and nematodes besides cold hardiness. Salt tolerant rootstocks have also been found possible in some progenies involving Cleopatra and Sunki mandarin and Rangpur lime.

Breeding Methods and Achievements

Introduction

Introduction of germplasm either from other countries or from one agro-climatic region to the other within the country has been one of the most potent improvement methods. The mandarin variety 'Santra' is known to have been grown in India for many centuries. It was introduced into the Central Provinces (now Maharashtra) by Ranghojee Bhonsal II from Aurangabad in eighteenth century. Tangerines, St. Michael Blood Orange and Large White Orange were imported and cultivated at Goojranwallah in Punjab during 1880. The present century has seen the introduction of a number of sweet orange varieties including Washington Navel, Valencia, Jaffa, Blood Red Malta and Tangerines. The first two were introduced from America and the others from the respective countries of their origin. Grape fruits were introduced from California and Florida, lemons from China and Malta from USA and Italy. 'Mosambi' seems to have been introduced in Nagpur during the beginning of the 20th century.

The introduction of 'Kinnow' mandarin (King x Willow leaf) in 1947 showed great promise in North India particularly Punjab. It was introduced in South India in 1958 and Punjab in 1959 and has performed extremely well in Punjab.

Clonal Selection

Exploitation of natural variability existing in a variety has resulted in the isolation of some promising clones in citrus.

- 'PKM 1 lime is a clonal selection from seedling progenies of kadayam type of Tirunelveli district of Tamil Nadu.
- 'Yuvaraj Blood Red' is a seedless and early maturing clonal selection from 'Blood Red' orange.
- 'Pramalini' and 'Vikaram', the two kagzi lime varieties were developed through clonal selection at Marathwada University.
- 'Chakradhar' is a thornless and seedless selection from Kagzi lime.

Hybridization

Hybridization is confronted with real problems in citrus improvement, both on scion as well as rootstocks because the long juvenile phase delays the assessment of the hybrids. Most of the cultivated varieties are highly polyembryonic, hence the crosses made using these as females result in very few weak hybrids, which are difficult to identify from nucellar seedlings. Electrophoretic techniques separating the isozymes of parents and hybrids may be of great value in scion breeding programme, as no morphological markers are available at present.

The heterozygous nature of the crop further leads to wide segregation. The problems are little less complicated of rootstock breeding where the commonly used disease resistant male parent *'Poncirus trifoliata'* has trifoliate leaves which is dominant over the monofoliate character (in other citrus varieties and all the hybrids), by which distinction of unifoliate nucellar seedlings could be easily made.

Hybridization Technique

The mature flower buds on the female parent are emasculated early in the morning on the day of opening and are bagged. The flowers to be used as male parent are bagged the previous evening. The next morning as the day warms up, the anthers dehisce releasing the pollen grains when these flowers can be plucked to pollinate the receptive stigmas of emasculated flowers. The pollinated flowers are bagged, opened after about a week and allowed to mature into ripe fruits. In some cases, especially when the trifoliate orange is used as male parent, difficulties are encountered as its flowering is over before other citrus varieties flower. Therefore, pollen has to be stored at low humidity and temperature.

Seeds from mature fruits are extracted and sown immediately in sterilized sand and soil mixture. When seedlings are about 15 cm high, hybrid seedlings are identified. Particularly those showing some morphological characters of male parent are marked while others are rejected. Electrophoresis methods can also be employed for identification of zygotic seedlings. Identification of hybrid seedlings having *P. trifoliata* as male parent is easily done by looking for trifoliate character. The hybrid seedlings are grown to mature trees in the field and the seedlings raised from the fruits are evaluated for resistance to various diseases, insect and pests, nematodes and for suitability as scion or rootstock.

Evaluation for Rootstock Purpose

Rootstock hybrids should have desirable attributes like high percentage of nucellar embryony, resistance to different diseases and nematodes. The selected hybrids are then tested with different scion varieties and compared with the commercial rootstock. Various plant and fruit characters yield and yield contributing characters are recorded.

Evaluation of Scion Hybrids

In the first round of evaluation, the zygotic seedlings are raised on suitable rootstock and observations on different vegetative and fruit characters are recorded. Meanwhile, the resistance to different diseases is also confirmed. Selected hybrids are tested on different rootstocks at different locations and compared with the commercial varieties.

Inter-generic and Intra-generic Hybrids

Inter-generic

Though inter-generic hybrids are rare in fruit plants, much success has been obtained in citrus.

1. Hybrids of Poncirus

Citrange: A group having the parentage of trifoliate orange (*Poncirus trifoliata*) and sweet orange (*C. sinensis*), the hybrids showed intermediate characters of the parents. The leaves are mainly trifoliate but unifoliate evergreen leaves are also observed in some plants. The fruits are juicy and flavoured. Some of the cultivars are Troyer, Carrizo, Morton, Etonia, Rusk, Coleman, *etc.*

In India, very little work has been done on citrus improvement through hybridization. At the PKV, Akola, hybridization work has been undertaken to evolve hybrids of kagzi lime. As a result, Hybrid 2, Hyrbid 4 and N52 were found resistant to canker. Breeding for improvement of citrus rootstock was initiated in 1972 at the Central Horticultural Experiment Station, Chethali, and IIHR, Bangalore. Trifoliate orange was used as a donor source for *Phytophthora* and citrus nematode resistance. Hybridization programme resulted in the production of 1183 hybrids from 16 different cross combinations. Of these, CRH-3, CRH-5 and CRH-41 resistant to citrus nematode have been evolved. A hybrid between Rangpur lime and trifoliate orange (Australia) having high resistance to nematodes and *Phytophthora*, and highly polyembryonic in nature is being evaluated for its suitability as rootstock for mandarin and sweet orange.

Mutation Breeding

Somatic mutations are common in citrus and through selection of the natural mutants; quite a few number of desirable clones have been obtained. The frequent occurrence of chimera may lead to clonal impurity and thus bud selection work in propagation becomes important for ensuring clonal purity. Selections of natural mutants have been successfully employed for seedlessness (Tangor), season of ripening (Satsuma, Navel), improvement of colour (Ray Ruby grapefruit), *etc.*

Besides natural mutations, many induced mutants have been developed in Citrus. For instance, 'Star Ruby' and 'Rio Red' varieties of grapefruit were developed in Texas, USA through X-ray and thermal neutron treatments of seeds of *cv.* 'Ruby Red' whose red flesh colour faded at harvest. In Japan, a few closely related clones of Satsuma mandarin with varied fruit colour and fruit ripening times were obtained through mutation. In USA also mutations had produced Satsuma seedling lines differing in productivity, fruit shape and the ripening time. The grapefruit clones like Thompson and Foster Pink arose as limb sports on white grapefruit. Gamma irradiation of seeds and bud woods performed in Orlando, Florida, resulted in seedless fruits on certain trees of seeded cultivars like Pineapple orange as well as Duncan and Foster grapefruit. In Israel, Shamouti trees of compact habit and early fruiting types and seedlessness have been developed in Eureka lemon through irradiation of bud wood with gamma rays.

Polyploidy Breeding

Most of the species and varieties of *Citrus* are diploids but occurrence of polyploidy has been reported in many cultivars. The Hongkong wild kumquats, *Fortunella hindsii* may have been the first reported tetraploid. Polyploidy breeding seems to offer prospects to obtain large sized fruit with dwarf plant types. Production of triploids by crossing tetraploid with diploids may be useful in obtaining seedless varieties. The seedless lime (*C. latifolia*) a triploid. Triploids have favorable characteristics and yield well but they are sterile. The development of triploid through breeding is very limited. Production of 3x is normally achieved by crossing of 4x with 2x which is often not feasible for want of sexual parents. The reciprocal cross ((2x) x (4x)) produces many tetraploid individuals. Polyploidy manipulation by crossing of tetraploids with diploids yielded some valuable triploid varieties like 'Oroblanco' and 'Melogold'. A large diversity of autotetraploid parents with desirable characters expressed in the progeny will be of high value to any citrus cultivar breeding program. Spontaneous autotetraploids occur among many polyembroyonic citrus varieties. Tetraploid trees of monoembryonic cultivars can be obtained by colchicine treatment. Triploids also, occasionally occur spontaneously as sexual seedlings. In most cases the egg provides the double chromosome number.

Biotechnology in Improvement of Citrus

Transformation of fruit species by biotechnological tools is a potential approach to develop disease resistant cultivars. Woody plants are known to be difficult to work *in- vitro* than herbaceous plants but citrus is exceptional. Though nucellar embryony in citrus is of great value for producing vigorous, uniform and virus free plants, it appears to be an obstacle in hybridization. In polyembryonic cultivars, the vigorous growth of nucellar embryos inhibits the growth of the zygotic embryo and causes its degeneration prior to seed maturation. Such abortive embryos can be rescued by tissue culture. Tissue culture has effectively been used in obtaining hybrid *Poncirus* plantlets from polyembryonic citrus cultivars. *Poncirus trifoliata* not only carries a genetic marker, but also possess resistance to tristeza, *Phytophthora,* nematode and cold stress. Inter-generic hybridization with the aid of cell/tissue culture offers possibility of incorporation of multiple desirable characters found in different genera for improvement of citrus root stocks and scion cultivars.

Cell and tissue culture and specially protoplast manipulations have effectively been explored in citrus improvement by regeneration of citrus trees from protoplast, somatic hybridization (cybridization) and organelle transfer. In an attempt to develop protoplast derived plants in the last one decade, Israel and Florida have shown protoplast system in a dozen genera and interestingly citrus is the only woody plants among them. Efficient protocols have been developed to obtain protoplasts with cell diversion capability from all major citrus cultivars and some of their wild relatives.

Important Species and Cultivars

Mandarin group (*Citrus reticulata*): Loose skinned orange, though mandarin and tangerine are names used more or less interchangeably to designate the whole group, tangerine is applied more strictly to those varieties which produce deep orange or scarlet fruits.

Calamondin (***C. madurensis***): Tanaka has recognized it as loose skinned orange group. It is very cold resistant for a true citrus fruit as hardy as Satsuma. Fruit colour is orange to deep orange, smooth and glossy surface, pitted shape, oblate, deep orange, and size small with flattened base having 7-10 segments.

Clementine (***Algerian tangerine***): It is a tangerine and is probably an accidental hybrid of the mandarin and sour orange which is considered to be originated in Algeria. Fruit colour deep orange, shape globose to elliptical, size-medium with depressed apex, rind thick, segments 8-12 adhered slightly. It is an early variety.

Cleoptara mandarin (***C. reshni***): It is originated in China. Plant is thornless with dense top. Fruits are produced singly or in clusters, fruit colour dark orange red, shape oblate flattened at both ends, size small and segments 12-15.

Coorg orange: It is an important variety of South India particularly in Coorg and Wynad tracts. Fruits are medium to large, bright orange colour, oblate to globose in shape, finely papillate and winkled, glossy, segments 9-11.

Dancy tangerine: In USA, the Dancy is the best known and highly prized of all the mandarin oranges. Tree large, nearly thornless and has upright growth. Fruit colour is deep orange red to scarlet, rind thin, loose, easily separable, segments 10-14. It is a late variety.

Deshi mandarin (Pathankot): This variety is mainly grown in Punjab hills. The trees are large with semi-upright growth habit and compact foliage and are spineless. Fruit are ovoid to sub-globose. Colour uniformly cadmium, surface pitted, semi-glossy and finely wrinkled, rind medium, adherence slight, segments 7-10.

King mandarin (*C. nobilis*): Swingle believed the king mandarin as a tangor, a hybrid between mandarin and sweet orange. King mandarin was first introduced from Cochin China to California in 1882. King mandarin is cultivated in Assam. This is a prolific bearer, frost resistant and produces high quality fruit.

Khasi mandarin: This is believed to be a hybrid between mandarin and sweet orange, and cultivated in Assam. It is a prolific bearer, frost resistant and produces high quality fruit.

Willow leaf mandarin (***C. deliciosa***): The tree is willow in growth, almost thornless, and fruits usually borne singly at the tip of slender branches. Fruit colour orange, surface smooth, glossy frequently slightly lobed, necked base, apex depressed, wrinkled, rind thin with 10-12 segments. It is an early variety.

Kinnow mandarin (King x Willow leaf mandarin): It is a first generation hybrid between the king and willow leaf mandarin and developed by H.B. Frost at the California Citrus Experiment station in 1915. It was introduced into Punjab from USA. Tree is vigorous, large, top erect, dense symmetrical with few scattered thorns. Fruit colour resembles of king, deep yellowish orange, surface, smooth, glossy, very shallow pitted, shape slightly oblate, size medium with flattened base, rind thin, peel tough and leathery, segment 9-10 easily separable, seed 12- 24. It is a late variety.

Nagpur santra: This variety occupies prime position in Indian market and is one of the finest mandarins grown in the world. It is also known as Ponkan. Tree is large, vigorous, and spineless with compact foliage. Fruit size is medium, cadmium colour, smooth surface, and glossy, rind thin, soft, and slightly adhered with 10-12 segments.

Satsuma orange (*C. unshiu*): It is a Japanese variety introduced into Florida in 1876. It is a frost resistant and useful breeding material. It is also resistant to canker, gummosis and scaly bark. Plant is thornless having spreading growth habit, orange fruit colour, rough surface, oblate to spherical shape, medium to large in size, thin and easily separable rind, flavour rich and seedless.

Temple mandarin: It is a hybrid between tangerine and sweet orange. Temple mandarin is most beautiful and highly flavored fruit of the citrus group. Tree is medium, thorny, spreading with deep orange to reddish fruit colour, rugose glossy surface, medium to large in size, depressed or nearly flat apex, loose rind, solid axis with 10-12 segments, orange pulp. It is late in maturity.

Lemon (*C. lemon*) Varieties of Lemon

Eureka: It is a seedling selection of Sicilian lemons. Tree is medium, spreading and thornless. Its fruit colour is lemon yellow, surface rugose, pitted, shape obovate, size medium, apex round, rind medium thin axis small, solid, segments 8-10, juice acidic with excellent flavor and quality. Eureka is a heavy yielder and begins bearing at early age. It has tendency of bearing in the terminal end of the shoot.

Lisbon: Its appearance and yield is superior to Eureka. It is resistant to frost, heat and high wind velocity. Tree is large and vigorous with spreading shoots. It has upright thorn growth, lemon yellow fruit colour, smooth surface, medium size, pitted rind, small axis, solid, 6-10 segments with 0-8 seeds.

Pant Lemon-1: Fruit size medium, juicy, heavy fruiting, tolerant to pests and diseases but could not become popular.

Vill Franca: It belongs to Eureka group and was introduced into Florida from Europe in 1875. Tree is vigorous, thorny, spreading, erect, fruit oval to oblong, size medium to large, colour bright lemon yellow, apex pointed, base rounded, rind thin, smooth, segments 8- 12, flesh fine grained, juice colourless, seed 25-30.

Meyer Lemon: Tree semi-dwarf, thornless, spreading, cold resistant, fruit colour light orange, surface smooth, finally pitted, shape oblate or oblong base rounded, rind thin, axis small, segments 8-10, seeds 8-12.

Acid Lime (*C. aurantifolia*): It is native of India and widely cultivated in the tropics. Tree medium sized, hardy, semi-vigorous, upright growth, thorny, fruit round to oblong, yellow apex rounded and slightly nippled, base round, rind thin, papery segments 8-10, seeds 8-10.

Varieties of Acid Lime

Vikram: It was developed at MAU, Parbhani, fruit medium size, heavy fruiting, fruit colour golden.

Pramalini: It was developed at MAU, Parbhani, high yielder, golden fruit colour, and tolerant to canker.

Sai Sarbati: Kagzi lime selection developed at Mahatma Phule Krishi Vidhyapeeth (MPKV), Rahuri, Maharashtra. Fruit surface smooth, fruits more uniform, good size, thin skin, high juice, TSS and acidity. It has high yield potential and tolerance to canker and tristeza.

Tahiti lime/Persian lime (*C. latifolia*): It is large fruited acid lime. The plants are large, spreading, cold resistant, thornless, fruit large in size, seedless triploid, and produce non-viable pollen. It is considered as hybrid between lime and lemon. Fruit colour orange yellow, smooth surface, segments 8-10. It is a late variety.

Rangpur lime (*C. limonia*): It is indigenous to India and is commonly used as root stock. Rangpur lime is mainly grown for home consumption and ornamental purpose. It is also known as Marmalade orange. It has loose rind, easily separable segments and pulp is light orange yellow.

Sweet lime (*C. limetoides*): Generally, sweet lime is grown as a root stock for its non acidic fruits.

Pummelo (*C. grandis*): It is native of Polinasia and Malaysia and commonly grown in South China. Fruit is pyriform, largest fruit size among citrus fruits, rind thick, juice is acid bitter, juice sacs easily separable. Seeds are monoembryonic. Fruits are of two types (a) elongated pear shaped with neck (b) oblate or globose, flattened and neckless. In India there is no improved cultivar except Nagpur Chakotra.

Varieties of Grape Fruit (*C. paradisi*)

Duncan: It was developed as chance seedling in Florida. It is the hardiest variety, fruit colour yellow, surface smooth, shape oblate to globose, size large, basal area depressed, apex round, rind medium thick, firm, axis medium in size, segments 12-14, and seeds 25-50.

Foster: It belongs to pink or red pulp group and originated as bud sport of Walters grape fruit by R.B. Foster in 1906-07. Fruit colour is light yellow, surface smooth, oblate or globose shape, size medium large, base rounded, apex round, rind medium thick, segments 12-14, seeds 2-5. It is a late cultivar.

Thompson: It is a bud sport of Marsh. Fruit colour light yellow, surface smooth, segments 10-12, and seeds 2-5.

Ruby: It belongs to pink or red pulp group. It is originated as bud sport from Thompson. Deep red colour which is uniformly distributed throughout pulp.

Citrus Hybrids

Hybrid of *Poncirus*

Citranges: It is hybrid between *Poncirus trifoliata* and *Citrus sinensis* and is hardy than sweet orange. It is used as dwarfing root stock for grapefruit, Satsuma, sweet orange and lemon.

Citrangors: A back cross hybrid between Citrange × Sweet orange.

Cicitranges: Citrange × Trifoliate orange.

Citrumelo: Trifoliate orange × grapefruits

Interspecific Hybrids

Limonage: *C. lemon* × *C. sinensis*

Tangors: *C. sinensis* × *C. reticulata*

Tangelo: *C. reticulata* × *C. paradisi*

Lemonimes: *C. lemon* × *C. aurantifolia*

Chapter 16

Grape

Botanical name: *Vitis vinifera* L.

Family: Vitaceae

Chromosome number: 2n = 2x = 38

History of Grape Breeding

European grape (*Vitis vinifera* L.) is considered to have originated primarily between Caspian and Black sea region. American grapes belonging to a large number *Euvitis* and *Muscadinia* species have originated in North America, referred to as 'Vine land'.

Botany

Genus *Vitis* has three types of flower, *i.e.,* male, female and hermaphrodite. *Vitis rotundifolia* is dioecious in nature. Further *Vitis* species had been originally dioecious sub dioecious but later on transformed to hermaphrodite. At present most cultivars of *Vitis vinifera* are hermaphrodite in nature. Inflorescence of cultivated grape is a cyme. Perfect flowers have 5 partly fused sepals and 5 petals joint at the top. The flowers have 5 stamens and pistil with 2 locule, short style and stigma. Time of anthesis is also influenced by environmental factors, hence varied reports with respect to anthesis are available. Peak time anthesis under North Indian conditions was between 7.00 to 9.00 am and South Indian condition was between 8.00 to 9.00 am.

Germplasm Resources

Field gene banks of grapes are maintained at Division of Fruits and Horticultural Technology, IARI, New Delhi, Indian Institute of Horticulture Research, Bangalore,

Ganesh Khind Botanical Garden, and National Research Centre, Grapes, Pune, *etc.* Further, 616 genotypes of grapes are maintained at IIHR, Bangalore.

Breeding Objectives

Objectives of breeding for grapes are:

- ☆ To develop early maturing, seedless and sweet cultivars for table purpose.
- ☆ To induce resistance to anthracnose, phylloxera and chaffer beetle.
- ☆ To develop varieties with medium vigour and productive basal bud, which can be trained on head or pandal system of training.

For the tropics the objectives of breeding should be:

- ☆ To develop high yielding and high quality varieties with increased fruitfulness of basal buds, less degree of apical dominance, suitability for different purpose such as table, raisin, wine and juice and resistance to diseases.
- ☆ To develop root stocks resistant to salinity, nematodes and drought.

Breeding Methods and Achievements

Introduction

Grapes are reported to have been introduced in Tropical India about 2600 years ago in 620 BC. Commercial cultivation did not start until the beginning of 20th Century. During 1930, Shree R.S. Pillay, identified Anab-e-Shahi from the collections of Nawab Baquer Ali Khan and subsequently its commercial cultivation picked up in South India. Bhokri and Cheema Sahebi in Maharashtra, Bhokri and Muscat Hamberg in Tamil Nadu and Bangalore Blue in Karnataka are the introductions.

The commercial varieties of grapes were introduced into India mostly by invaders of Iran and Afghanistan. Muhammed Bin Tughlaq introduced Bhokri, Fakhri and Sahebi cultivars in Aurangabad (Daulatabad) in 1338. Large-scale introduction in a planned manner were initiated at Lyallpur as early as 1928, when S.B.S. Lal Singh, was Head of Department of Horticulture, introduced as many as 116 grapes varieties from different grape growing countries. The earlier promising introductions include, Thompson Seedless, Perlette, Beauty Seedless, from USA, Kishmish Beli and Kishmish Charni from USSR.

The cultivars like Ruby Seedless, Gordo Blano, Reisling, MS 18-55, MS 19-77, MS 16- 2, Wortly Hall hybrids from Australia, Totlocha from Brazil Flame Seedling 1281, Dogridge, Pride, Dixie, Wedor and Black Cornith-2 from USA, Surnak Kitabiskij, Pozdrijwir and Shirajx-6 from USSR, Malvasiafina (Douro), Boal De Alicante, Tinta Deira Preta, Jampal, Tinta Roriz, from Portugal and 0912 Horizon (SW), 0913 Leon Millet, Foch and 0912 Swanson Red from Canada for wine, raisin and table purposes have been introduced and are under evaluation. Further, number of *Vitis* sp. have been introduced for resistance to biotic and abiotic stresses, *e.g., V. gigas,*

V. caribea, *V. munsoniana*, *V. smalliana*, *V. cineraria*, *V. shuttleworthi*, *V. arizonica* and *V. monocola* from USA.

Selection

Open-pollinated seedling segregates for a large number of characters and hence the population of seedlings from open pollinated seeds is a potential source for selection of desirable type, *e.g.*, Cheema Sahebi (Sel-7), Selection-49. Some promising seedlings from open-pollinated population of Pandhari Sahebi and Kabul Monukka were also selected.

Clonal selection is also one of the methods of fruit improvement. Due to natural mutation in existing cultivars considerable variation occurs between individuals that help in varietal improvement through clonal selection. The promising clonal selections of grapes are as follows:

Cultivars	*Clonal Parent*	*Characteristics*
Tas-A-Ganesh	Thompson Seedless	Developed by Mr. Arue of Borgoan in Sangli district of Maharashtra. Its berries are quite elongated and respond to GA_3 treatment.
Rao Sahebi	Cheema Sahebi (Sel-7)	Isolated by Rao Saheb Kadlag of Sangamner in Nasik district of Maharashtra. Fruits have longer berries with stronger attachment to rachis which is a major problem in Cheema Sahebi.
Sonaka and Manik Champa	Thompson Seedless	Sonaka has much elongated berries as compared to Tas-A-Ganesh. It gives better response to GA_3.
Dilkhus	Anab-e-Shahi	Selection was made at Hyderabad. It produces golden yellow elongated seeded berries in attractive bunches. The yield potential is almost same as in parent.

Selection Made by Institutes

- **Pusa Seedless from Thompson Seedless:** Developed at IARI, New Delhi. It differs from the parent in respect of having more elongated berries. Vine vigorous and heavy yielding: TSS 22-24 per cent, acidity 0.77 per cent and juice content 65 per cent. It ripens in the middle of June.
- **HS 37-6 from Perlette:** Developed at HAU, Hissar. This cultivar is 15 days earlier in maturity than the parent.

Hybridization

Grapes are highly heterozygous and are propagated asexually at commercial scale. Inbreeding results in rapid loss of vigour and fertility of vine, even in first generation. The crossing of unrelated parents with good combining ability followed by raising a large number of hybrid seedlings in each combination and vigorous selection may result in good ideotype of commercial use.

In India, hybridization work was started in 1958 at IARI, New Delhi. The purpose of hybridization at IARI, New Delhi was to develop early maturing, high yielding, and better quality seedless varieties with resistant to biotic stresses. However, IIHR, Bangalore, started breeding programme in 1968, with objective to develop superior varieties for table, raisins, wine and juice, On the basis of types of parent used, it can be grouped into two (a) Interspecific or Intergeneric hybridization (b) Interspecific or intervarietal hybridization.

Inter-specific/Inter-generic Hybridization

Muscadinia is a rich source of resistance to diseases and pests and also possesses a unique and delightful flavor and aroma. The crosses between *Vitis* and *Muscadinia* which differ in chromosome number are made with difficulty, but most of the resulting hybrids remain sterile. The pollen of *M. rotundifolia* will fertilize the egg cell of *V. vinifera* but the reciprocal cross is less successful. Partly fertile F_1 hybrids (2x=39) can cross reciprocally between themselves or with *V. vinifera* x *M. rotundifolia* which have been further improved by back crossing with *V. vinifera*, resulting in some fertile vines that produce acceptable quality table grapes. Crossing within *Muscadinia* has given outstanding self fertile cultivars like Tarheel (*M. rotundifolia* x *M. munsoniana*), South Land, Magron, Regale (cold hardy) Sterling (cold hardy) and Triumph (bronze cloured berry weighing 7.9 g). Telki 5A (*V. berlendieri* x *V. riparia*) highly resistant to *Phylloxera*, tolerant to lime soils and moderately resistant to nematodes, Harmony (1613 x *V. champini* planchon *cv.* Dogridge) has been developed as a result of inter-specific hybridization.

Inter-varietal Hybridization

A few promising hybrids identified through inter-varietal hybridization at IARI were, Hybrid 62-37 (Hur x Pusa Seedless), H62-65 (Hur x Pusa Seedless), H-62-20 (Hur x Black Hamburg) H-62-67(Hur x Bharat Early), H-63-10 (Bhokri x Pearl of Casaba), H- 63-32 (Bhokri x Pearl of Casaba). In 1996, cultivars Pusa Navrang (Madeleine Angevine x Rubi Red and in 1997 Pusa Urvashi (Hur x Beauty Seedless) were released from IARI, New Delhi. The promising hybrids developed at IIHR, Bangalore were Arkawati (Black Champa x Thomson Seedless),Arka Kanchan (Anab-e-Shahi x Queen of the Vine Yards), Arka Shyam (Bangalore Blue x Black Champa), Arka Hans (Bangalore Blue x Anab-e-Shahi), Arka Chitra (Angur Kalan x Anab-Shahi), Arka Krishna (Black Champa x Thompson Seedless), Arka Majestic (Angur Kalan x Black Champa), Arka Neelmani (Black champa x Thompson Seedless), Arka Soma (Anab-e-Shahi x Queen of the Vine Yards), Arka Trishna (Bangalore Blue x Convert Large Black), Arka Shweta syn., Shweta Seedless (Anab-e-Shahi x Thompson Seedless).

Hybridization Technique

It includes the choices of parents, emasculation, pollination, shortening of breeding cycle for early assessment, growing of hybrid seeds and planting in the field for assessment and selection.

(i) Choice of Parents

In order to incorporate the desirable characters of one cultivar into other through hybridization, the knowledge of inheritance pattern and general and specific combining ability of the cultivars is very essential for making choice of parents in restricting the cross-combination and more seedlings population for better selection. The viability and germination ability of the hybrid seeds are also important factors in deciding the parents to be used in hybridization. It has been found that in some cultivars when used as female parents or selfed, the seed germination is poor and some time do not germinate, *e.g.,* Cordinal. If such cultivars are required in hybridization, they should be used as male parents in order to induce seedlessness in the progeny. It would be better to select a variety having high seed index as female parents. Cultivar Angoor Kalan can also be used as female parent for earliness, seedlessness and good quality, but for the same purpose cultivars Beauty Seedless, Perlette and Pusa Seedless should be used as male parents.

(ii) Emasculation and Pollination

Emasculation of small flowers of grapes is a tedious job. Since the grape is self fertile emasculation is most essential for making desired crosses. Use of reflexed stamens and functionally female cultivars like Hur, Angoor Kalan, Banquiabyad, Katta, Kurgan as female parents can help in eliminating the tedious task of emasculation. Iyer and Randhawa (1966) reported that aqueous solutions of mallic hydrazide (MH) at 400 to 750 ppm, 2, 3, 4 Tri iodo – benzoic acid (TIBA) at 400 to 500 ppm and 1, 2 dichloro-iso-butyrate (FW-450) at 0.30 per cent applied twice to 13 to 15 days old inflorescence induced pollen sterility. When emasculation is completed the emasculated bunches are bagged and pollinated with desired male parents very next day.

Mutation Breeding

Mutation breeding may be attempted as a complementary tool in grape breeding for one or more important characters, without altering the whole genetic setup. The important mutagens used in grape breeding are physical mutagens (X-ray and Gamma rays) and chemical mutagens (Ethyl Methane Sulphonate (EMS), N-Nitroso-N-Methyl Urethane (NMUT) and N-Nitrose-N-Methyl-Urea (NMU).

Further, induced mutations have resulted in a few improved varieties, New Perlette (Loose Perlette) with comparatively loose bunch has been evolved with X-rays (2.5 KR) treatment on Perlette Self thinning property of New Perlette is a result of meiotic irregularities caused by chromosomal translocation. Red Niagara having red fruit from Niagara and Robin Cardinal an early maturing variety from Cardinal are other important induced mutants in grapes.

Polyploidy Breeding

Polyploidy breeding has immense importance in the improvement of table grape. The chief benefit from polyploidy is the increase in berry size. However, autotetraploids are found to be considerably sterile and are less productive than the parents. The crossing of diploid with induced tetraploids may help in evolving new triploid seedless grapes. The triploids are highly sterile. All tetraploids even between infertile species have been more desirable as commercial varieties. Colchicine is generally used as an aqueous solution of 0.25-5.0 per cent with 5-10 per cent glycerine to induce polyploidy. Marvel Seedless from Delight, Early Niable (Campbell x Niagra), Lonetto, Early Giant from Campbell, Muscat Common Hall from Muscat Alexandria, Black King from Campbell, Wallis Giant from Concord, Case from Sultana, *etc.* are important examples of polyploidy.

Biotechnological Tools Embryo Rescue Technique

Seedlessness is a desirable character for table and raisin grapes. Inheritance of seedlessness is postulated to depend on two complementary recessive genes and only about 7.5 per cent of the total progeny from crosses between Seeded x Seedless grapes produced fruits without noticeable seed traces. The embryo rescues theoretically increases the proportion of seedless progeny as it makes possible to cross two seedless varieties. Ovules are excised before abortion and are cultured on either filter paper in liquid medium or solid medium.

Genetic Engineering/Plant Transformation

Some encouraging preliminary results have been obtained on *Agrobacterium* mediated transformation of grape vines. But the production of genetically transformed grape vines which express a marker gene is yet to be reported.

Protoplast Culture

Protoplasts are of great importance as tool for genetic amelioration and somatic hybridization. But regeneration of grape vines from protoplasts has not yet been successful.

Anther Culture

Anther culture can result into haploid grape vines which can then be developed into homozygous diploids by doubling chromosomes. These homozygous diploids will be very useful for producing F_1 hybrids and for making genetic studies. But there is low success rate of regeneration of grape vines from anther and only one case of haploid has been reported in grape.

New Cultivars of Grape

Arkawati (Black Champa x Thompson Seedless): Bunch is medium in size, yellowish green berry, sweet, TSS 22-25 per cent, seedless berry, suitable for raisin making, fresh table use and making good quality dry and white table and dessert wine. It was released in 1980.

Arka Chitra (Angur Kalan x Anab-e-Shahi): Released in 1994, this is tolerant to powdery mildew, moderately vigorous vine, good yield potential (34kg/vine), Bunch is well filled, medium to large (310g), berry very attractive; golden yellow with pink blush, slightly elongated, large (3.18 g), sweet, TSS 20-21°Brix, acidity 0.4-0.6 per cent, suitable for table purpose, all the buds are fruitful, suitable for head system of training and gives two crops in a year.

Arka Hans (Bangalore Blue x Anab-e-Shahi): Released in 1980, bunch is medium in size, berry yellowish green, sweet, TSS 18- 21°Brix, having foxy flavour seeded cultivar, suitable for making quality wine, resistant to anthracnose.

Arka Kanchan (Anab-e-Shahi x Queen of the Vineyards): Bunch is large, golden yellow colour berry, ellipsoidal to ovoid, sweet, TSS 17- 20°Brix, having muscat flavour, seeded cultivar, suitable for fresh table use and dry white table and dessert wines, released in 1980.

Arka Krishna (Black Champa x Thompson Seedless): Vigorous vine, good yield potential (30kg/vine), Bunch well filled berry, dark colored, seedless, sweet TSS 20-21°Brix, acidity 0.6-0.7 per cent, suitable for head system of training and gives two crops in a year, suitable for juice making, released in 1980.

Arka Majestic (Angur Kalan x Black Champa): Released in 1994, vine is vigorous, high yield potential (34 kg/vine), Bunch is well filled, medium to large (370g) berry deep tan colour, sweet, TSS 18-20°Brix, acidity 0.4-6 per cent all the buds are fruitful, suitable for head system of training, it gives two crops in a year tolerant to anthracnose.

Arka Neelamani (Black Champa x Thompson Seedless): Vigorous vine with high yield potential (25 kg/vine), bunch well filled, sweet, TSS 20-22° Brix, suitable for head system of training, it is good for table purpose and making red desert wine, tolerant to anthracnose, released in 1980.

Arka Shyam (Bangalore Blue x Black Champa): It was released in 1980, bunch is medium in size, berry bluish black, spherical to obovoid, sweet, TSS 20-25°Brix, having mild foxy flavour and seeded, it is good for fresh table use and making dry table, dessert wines and juice, resistant to anthracnose disease.

Arka Soma (Anab-e-Shahi x Queen of the Vineyards): This variety was released in 1994, vine is vigorous with heavy yield potential (36 kg/vine), bunch is well filled large (410 g) berry greenish yellow, round to ovoid large (3.8 g), sweet, TSS 20-21°Brix, acidity 0.5 per cent having muscat flavor, gives two crops in a year, suitable for good white dessert wine.

Arka Trishna (Bangalore Blue x Convent Large Black): It was released in 1994, it is an improvement over variety Bangalore Blue, vine is vigorous, having high yield potential medium bunch, well filled, very sweet, TSS 22- 23°Brix, acidity 0.3- 0.4 per cent it is male sterile hybrid, good for wine making suitable for head system of training, resistant to anthracnose and tolerant to downy mildew.

Arka Shweta or Shweta Seedless (Anab-e-Shahi x Thompson Seedless): It was released in 1994, moderately vigorous vine, yield potential is about 28 kg/vine, bunch is medium, it responds to GA_3 application for berry thinning and enlargement, berry seedless, sweet, TSS 18-19°Brix, acidity 0.5-0.6 per cent, berry greenish yellow.

Digrasset: This variety was collected at ARI (MACS), Pune from the grape germplasm collection maintained at Ganesh Khind Botanical Garden, Pune in 1976. It is a clone of *Vitis champini*, vine shows vigorous, spreading and prostrate growth having deep root system. It remains dormant during winter season after October pruning and again grows in February-March under climatic conditions of Maharashtra. This is a potential root stock for growing grape under saline and drought conditions.

Pusa Navrang (Madeleine Angevine x Rubired): It was released in 1996, it is basal bearing, tenturier (peel and pulp both coloured), seeded cultivar, it is early maturing, suitable for making coloured juice and wine, bunch is loose and medium TSS 19°Brix, resistant to anthracnose.

Pusa Urvashi (Hur x Beauty Seedless): It was released in 1997, it comes in early group, bunch is medium, and berry is greenish yellow, TSS 20-22 °Brix.

Chapter 17

Pomegranate

Botanical name: *Punica granatum* L.

Family: Punicaceae

Chromosome number: 2n = 2x =18

History of Pomegranate Breeding

Pomegranate is native of Iran and cultivated extensively in the Mediterranean countries like Spain, Morocco, Egypt, Iran, Afghanistan and Baluchistan. It is also grown to some extent in Burma, China, Japan, USA, USSR and India.

Botany

Inflorescence of pomegranate has been reported to be cyme. There are two flowering season in North India whereas three flowering season in South India. There are three types of flowers present on same plant of pomegranate, *i.e.,* male, hermaphrodite and intermediate. Ovary of male flower is rudimentary whereas that of intermediate flowers are degenerating type. If fruit set takes place in such flowers they may drop before reaching to maturity even if some fruits reach maturity that become misshaped. Heterostyly is also common in pomegranate in this case hermaphrodite flowers are pin eyed and male flower are thrumb type.

Germplasm Resources

Being cross-pollinated crop, a lot of variability exists in seedling populations, which can be utilized in further improvement programme. At present, 150 genotypes of pomegranate have been maintained at Central Institute of Arid Horticulture, Bikaner. Out of these genotypes, 55 are deciduous and rests 95 are evergreen in nature. Field gene banks of pomegranate are maintained at Abohar, Rahuri, Bikaner, Bangalore, Allahabad, Jodhpur and Ludhiana.

Breeding Objectives

- ☆ To develop suitable types which produce small soft seeds with attractive red (pink) aril.
- ☆ To develop easily manageable upright growth habit of the tree.
- ☆ To develop thornlessness in the twigs, a desirable character as it helps in cultural management of the tree.
- ☆ To develop varieties resistant to fruit borer (*Virachola isocrates*) and fruit rot (*Phomopsis spp.*).
- ☆ To develop varieties free from fruit cracking, aril blackening.
- ☆ Identification and development of suitable varieties for cold arid region.
- ☆ Varieties with longer storage life.

Breeding Methods and Achievements

Introduction

Some important cultivars including soft seeded, dark red grained types, *viz.*, Wonderful from the USA, A. Males, Be Hastah, A. Alah, A. Agha, Mohammed Ali, A Post Sephid Sirin from Iran and Ranninj G-1-8-23, Rannyij G-1-3-34, Chereny,Gulsha Red, JG-1-8-7 from USSR and few cultivars from Tunisia have been introduced. At Hissar, cultivar Shirin Anar and Russian Seedling were found resistant to bacterial leaf spot. A pomegranate line of Iranian origin has been indentified at Rahuri which has dark pink arils, soft seeds and high TSS.

Selection

Many pomegranate types cultivated in India are of seedling origin. They offer a wide range of variability with respect to shape and size of fruits, mellowness of seeds, aril colour, rind colour, sweetness and acidity. On the basis of yield and physico-chemical characters of fruits, number of cultivars have been recommended for commercial cultivation in different states of India, *viz.*, Ganesh, G-137, P-23, P-26 and Muscat in Maharashtra, Bassein Seedless, Jyothi and Madhugiri in Karnataka, Dholka in Gujarat, Jalore Seedless, Jodhpur Red, and Jodhpuri White in Rajasthan and Kabul Red, Vellodu, Yercaud 1, and CO-1 in Tamil Nadu. Two ornamental types (Japanese Dwarf and Double flower giving red, yellow and white flowers) are planted in the ornamental gardens. Due to considerable variability and their adaptability to existing agro-climatic conditions, selection of superior genotypes will be the best approach to get desirable ideotypes. The cultivar GBG-1 is a selection from open pollinated population of Alandi in 1932. The name Ganesh was given in 1970. Five Muskat types, namely P-13, P-16, P-23, P-26 and SK-1 were identified by Naik (1975). Further, P-23 and P-26 were released in 1986 for commercial cultivation by MPKV, Rahuri. At the University of Agricultural Sciences, Bangalore as a result of evaluation of seedling populations raised from Bassein Seedless and Dholka Varieties, GKVK-1 now named as Jyothi was released. At Coimbatore self seeded selection CO-1 was identified.

Clonal Selection

G-137 is a superior clonal selection over Ganesh, other clones are also superior, *i.e.*, G-107, G-132, G-133. A clone Acc. No. 455 which has been renamed as Yercaud-1 and released for commercial cultivation in Tamil Nadu.

Hybridization

In order to incorporate blood red colour of Russian types into Ganesh, several crosses were made at Rahuri in 1976. Out of 122 F_1 hybrids, seven had deep red aril colour but the seeds were hard and inferior in taste than Ganesh. A promosing line from the F_2 population (No. 61) combining desirable quality attributes has been released by the name Mridula (Ganesh x Gulsha Rose Pink).

Mutation

Use of physical (X-rays) and chemical mutagens (N, N-dimethyl N-nitrosourea) may help in the development of the superior cultivar of soft seeded types.

Biotechnological Tools

Attempts have been made to regenerate the plant by using leaf and shoot tip explants. Enzyme based marker was also used to identify the genetic variability among the existing genotypes. Somatic embryogenesis was also practiced by using petal as explant. Important characteristics of some promising selections raised from open pollinated fruit of F_1 hybrids of pomegranate.

Important Characteristics of some Promising Selections

Characteristics	*Sel-5 (Ganesh x Shirin Anar)*	*Sel-130 (Ganesh x Gulsha Rose Pink)*	*Sel-303 (Ganesh x Gulsha Red)*
Fruit colour	Apple red	Greenish brown	Yellowish brown
Fruit weight (g)	130	107	140
Fruit size LxB (cm)	6.9x6.5	6.1x6.5	5.5x6.8
Aril colour	Blood red	Dark red	Blood red
Mellowness of seeds	Soft	Soft	Soft
Taste	Sweet	Sweet	Sweet
No. of grains/100g	58	94	45
Grain peel ratio	1.17	1.17	1.80
Juice colour	Dark red	Blood red	Blood red
Juice (per cent)	80	80	80
TSS (per cent)	18.4	15.8	19.0
Acidity (per cent)	0.64	0.80	0.64

Cultivars of Pomegranate grown in different States

Name of the States	*Cultivars*
Rajasthan	Jalore Seedless, Jodhpur Red, Jodhpuri White.
Haryana	Ganesh, Muskat Red, Paper Shell.
Gujarat	Dholka, Muskat Red, Paper Shell.
Maharashtra	Ganesh, G137, P23, P26, Muskat Mridula
Karnataka	Bassein Seedless, Jyothi, Paper Shell, Madhugiri.
Tamil Nadu	CO-1, Yercaud, Vellodu, Kabul Red.

Description of Important Cultivars

Alandi: Also known as Vidaki, medium fruit size, fleshy testa, blood red or deep pink with sweet, slightly acidic juice with hard seeds.

Dholka: Large fruit size, greenish white rind, fleshy testa, pinkish white or whitish with sweet juice, soft seeds and acidic juice.

Kabul: Large fruit size, rind deep red mixed with pale yellow, thick, fleshy testa dark red, slightly bitter juice.

Kandhari: Fruit large in size, rind deep red, fleshy testa, blood red or deep pink with sweet, slightly acidic juice, hard seeds.

Muskat Red: Fruit small to medium in size, rind somewhat thick, fleshy testa with moderately sweet juice, seeds are semi hard.

Paper Shell: Fruit medium in size, rind thick, fleshy testa, reddish pink with sweet juice and soft seed.

Spanish Ruby: Fruit small to medium in size, rind thin, fleshy testa rose coloured, soft seed.

Ganesh: Prolific bearer, medium fruit size, soft seeds, sweet in taste.

Jyothi: Also known as GKVK-1, attractive yellowish red fruit colour, medium fruit size, red aril colour and soft seeds.

Vellodu: Fruit medium to large in size, rind moderately thick, fleshy testa, juicy, seed moderately hard.

Poona: Fruit large in size, fleshy testa, deep scarlet or pink and red.

Bedana: Fruit medium to large in size, rind brownish or whitish, fleshy testa, pinkish white with sweet juice and soft seeds.

Bhagwa: It is developed by MPKV, Rahuri. It is tolerant to thrips and mites, it is free from blackening of arils and there is no incidence of fruit cracking. Fruits have cherry red bold aril.

Phule Arakta: It is also developed by MPKV, Rahuri. Plant is heavy yielder with bigger fruits and sweet soft seed. It is less susceptible to fruit spots and thrips.

Chapter 18

Pineapple

Botanical name: *Ananas comosus* L.

Family: Bromeliaceae

Chromosome number: 2n = 2x = 50

History of Pineapple Breeding

Pineapple is believed to be originated in Brazil. The wild Brazilian pineapple (*Ananas microstachys* Lindle) is considered as ancestor of cultivated pineapple. It reached India during 1548. The generic name "Ananas" is derived from the Indian "Nana".

Botany

Pineapple plant generally flowers after attainment of certain vegetative growth and it is attained in 11-12 months after planting and formation of at least 40 leaves. Pineapple plant produces one fruits in its life. *Ananas comosus* is the only self-incompatible species in genus, whereas other wild species and Pseudananas are partially self fertile. Pineapple showed gametophytic S-allele types of incompatibility and induction of polyploidy does not breakdown the incompatibility system. After completion of the vegetative growth terminal bud is terminated to peduncle which bears inflorescence. In pineapple fruit develop parthenocarpically. It is not known whether pollination is required to initiate the fruit development.

Inheritance Pattern

According to Collins, 1960 spineless form in which spines are restricted to top few inches of the leaf are dominant to spiny wild type. Further many wild cultivars possess spiny leaves. *Ananas ananassoides* is also a source of disease resisteance.

The cultivar Pernambuco is donor for good flavour and aroma, tender non-fibrous juicy fruits, early fruiting, resistant to heart and root rot. Queen can be donor for crisp, non-fibrous deep yellow flesh fruits and early ripening.

Red Spanish is good source of vigour, resistance to wilt, heart and root rot. Singapore Spanish can be good donor for square. Shouldered fruits with golden yellow flesh and wild species are good source of vigour, resistance to various disease and pests, spiny tip and spiny characters are the phenotypic expression of single pair of alleles, with spiny tip being dominant. Homozygous SS and heterozygous S produce spiny tips, recessive and may give rise to spiny plants progeny. The piping and non-piping characters are controlled by another non-linked pair of alleles with the gene P (piping) being epistatic to S and s. The homozygous pp genotype produces pronounced piping than Pp genotypes. Frequent mutations of S and s occur. *Ananas erectifolius* is having S^e gene of smooth tip leaves.

Germplasm Resources

In India, much attention has not been given in the development of field gene banks. However, commercial cultivars are maintained at BCKV, Kalyani, Agriculture Research Station, Kovvur (APAU), Department of Horticulture, College of Agriculture, Jorhat, Regional Research Station: Diphu (Assam Agricultural University), Department of Horticulture, College of Horticulture, Navsari (GAU), College of Horticulture at Vellanikara, Trichur. A list of 135 varieties was published as early as in 1935 by Johnson, although some of them were found synonymous.

Breeding Objectives

- To develop high yielding, early maturing varieties with wider geographical adaptability.
- Plant should be hardy, vigorous, capacity to produce good ratoon crop, leaves should be spineless.
- Fruit stalk should be short and strong.
- There should be flat eyes and small cones.
- Development of varieties resistant to biotic and abiotic stresses (*e.g.,* multiple crown, fasciations, wilt, heart rot, root rot and nematode).

Breeding Methods and Achievements

Selection

Most of the cultivars/varieties of pineapple were developed by simple selection of mutant clones within cultivars and by hybridization between cultivars followed by selection from the highly heterozygous progeny. Selection from the Singapore Spanish population in Malaysia had led to a new cultivar, "Masmerah". It is more vigorous, possessing more leaves, and more erect and bears heavier fruit than the parent cultivar. Several such selections have been made in different pineapple growing areas.

Hybridization

A cross between Red Spanish and Cayenne has led to the development of a new hybrid PR-1-67 in Puerto Rico. This hybrid shows better plant vigour and resistance to wilt disease. However, self fertile somatic mutants obtained from cultivar Cayenne show a loss of vigour on selfing and heterosis on crossing. A hybrid (H-7) has been produced by crossing Valera Monendi x Kew. This hybrid produces large fruits, individually weighing on an average 3.0-3.5 kg.

Mutation

Induction of mutation in this crop seems to be quite feasible. However, due to wide natural variation, limited attempts have been made for induced mutations. In Kerala, irradiation of plants of cultivar Kew and Mauritius led to growth retardation and premature suckers. The induction of self fertile mutants by X-rays irradiation of pollen during meiosis. Several morphological mutations were found when 1.0 to 1.5 month old detached slips were treated with chemical mutagens like Ethyl imine-(EI), N-Nitroso-N-Methyl Urethane (NMU) and Diethyl Sulphate (DES). One mutant produced spineless plants from cultivar Queen and was economically significant.

Biotechnological Tools

Attempts have been made for rapid multiplication of the plants through micro-propagation by using different kinds of explants, *i.e.,* leaf base, shoot base, excised lateral buds, meristem tips from crown, *etc.* In the crosses where fertilization fails due to incompatibility, embryo culture technique can help to rescue the hybrid. The genetic transformation of pineapple clones has been attempted with the objective to acquire ability to introduce desirable genes.

Major Group and Varieties of Pineapple

Groups	*Cultivars*	*Important Characteristics*
Abacaxi (Brazilian)	Abakka, Amarella, Papelon, Pina Valera, Sugar Loaf, Venezolana, Vermelho, Yupi	Fruit conical in shape, weighing about 1.2 to 1.5kg, yellow rind, pale yellow or white flesh, sweet, tender, and juicy, leaves spiny, disease resistant, grown for fresh domestic consumption.
Cayenne	Boron, Baraonne de Rothschild, Champaka, Cayenne Lisse, Emeralda, Gautemalan, Giant Kew, Hilo Cayenne, Kew, Rothschild, Smooth, St. Micheal Smooth Cayenne, Typhone.	Fruit shape is cylindrical with a slight upward tapering and flat eyes, most suitable for canning fruit weight is about 1.8 to 3.0 kg. Colour of rind is dark orange and flesh is pale yellow, sweet, mildly acid with low fiber and a tender juicy texture, leaves smooth with few spines near the tip, highly susceptible to mealy bug and wilt, suitable for export.
Vaipure	Bumanguesa, Legrija, Maipure, Mar quita, Monte Lirio, Perolera, Plamba de, Rondon	It is sweeter than the Cayenne, aromatic, fibrous but tender and very juicy, leaves completely smooth, grown for fresh consumption, fruit ovoid to cylindrical in shape, fruit weight is about 0.8-2.9 kg, rind colour is yellow to dark orange or red, flesh is white or deep yellow.

Groups	Cultivars	Important Characteristics
Queen	Alexandria, Lakhat, James, Jhaldhup, Mac Gregor, Mauritius, Netal, Queen, Ripley, Victoria, Z. Queen	Fruit shape is conical, weighing about 0.5 to 1.12 kg rind is yellow, flesh is deep yellow, less acidic than Cayenne, sweet, low in fiber, spiny leaves, highly resistant to diseases than the Cayenne.
Spanish	Betek, Cabezona, Castilla, Espanola Roja, Gandol, Green Selangor, Masmerah, Nangka, PRI- 67, PRI-56, Red Spanish, Singapore Spanish.	Fruit shape is globose, fruit weight is 0.9-1.8 kg rind deep reddish, flesh pale yellow to white with spicy acid taste, fibrous texture, leaves spiny, resistant to mealy bug and wilt, susceptible to gummosis, suitable for export and fresh consumption.

Chapter 19

Sapota

Botanical name: *Achras zapota*

Family: Sapotaceae

Chromosome number: 2n = 2x = 26

History of Sapota Breeding

It is a wind pollinated crop. Flowers are protogyny and the stigma grows out of the bud about two days before anthesis. Flowers open between 4.00-4.30 am. Anthers dehisce between 8.00-10.00 pm. The flowers keep fresh for nearly two days. The stigma is found to be receptive two days before opening and continues to be like that up to 12 hours after opening. Peak receptivity is between 8.00-10.00 am. The total time taken from fruit set to maturity is 10-12 months under North Indian conditions but in Tamil Nadu it takes only 4-5 months.

Flowers are emasculated and bagged before 4.00-5.00 pm and well before the stigma protrudes out of the bud. The actual procedure consists of making a circular incision around the flower bud with sharp knife or blade, so that 2/3rd of the upper floral cup is removed including the portions of calyx, corolla and epipetalous stamens. The style is left in position in remaining 1/3rd of the floral cup. Stamens from male parent, which should shed their pollen in the early hours of next day, are collected in the previous day evening and kept overnight in a petri dish. These are used to pollinate the receptive stigma of the emasculated flower between 8.00-10.00 am in the next day.

Breeding Objectives

The main emphasis on breeding of sapota are to develop dwarf stature trees with precocity in bearing, high yield and high keeping quality of less seeded fruits with less latex.

Breeding Methods and Achievements

Clonal Selection

Numbers of varieties like Cricket Ball, Kirthi Barthi, Oval, Thagarampudi, Badami, Baramasi and Guthi exhibit natural variability. Exploration of this natural variability by clonal selection is an accepted method of breeding in sapota.

CO-2: Developed at Tamil Nadu Agricultural University, Coimbatore is a clonal selection from Baramasi. It is a high yielder; seeds are less in number and small sized (2-3).

PKM-1: Developed at Horticultural College and Research Institute, Periyakulam (TNAU) is also a clonal selection from Guthi. It is a dwarf, high yielding (3600 fruits/ tree/year), almost bearing throughout year.

PKM-4: A clonal selection from open-pollinated seed of PKM-1. It has spindle shaped fruits suitable for dry flakes production. Pulp is attractive with light pinkish honey brown colour, crisp and sweet flesh (TSS 24° Brix).

Hybridization

Tamil Nadu Agricultural University, Coimbatore has developed four hybrids so far.

CO-1: It is a hybrid between Cricket Ball and Oval. This variety is superior to either of the parents. The fruits are long oval (egg shaped), medium in size with a mean fruit weight of 125 g. The flesh is granular in texture and reddish brown in colour, taste being very sweet with a TSS of 18° Brix.

CO-3: It is hybrid between Cricket Ball and Vavivalasa. Fruits are oblong-ovate in shape pleasantly flavoured, very sweet with a T.S.S of 24.2° Brix. The average yield of the tree is 157 kg as compared to only 101.32 kg and 109.5 kg in CO-1 and CO-2 respectively. The stature of the tree is more upright and compact, suitable for high density planting at a spacing of 5-6 m either way instead of the conventional spacing of 8m x 8m.

PKM-2: It is a hybrid between Guthi and Kirthi Barthi developed at 'Horticultural College and Research Institute, Periyakulam (TNAU). A high yielder with a performance of 1500 to 2000 fruits per tree per year weighing 80 to 100 kg. Fruits are bigger in size and oblong to oval shaped. The average fruit weight is 95g, TSS ranges from 25 to 27° Brix.

PKM-3: It is a hybrid between Guthi and Cricket Ball. It has vertical growth habit and hence lends itself for high density planting. Trees bear big sized fruits with oval shape and have cluster-bearing habit. The fruit yield is 14 tones per hectare.

DHS-1: A hybrid between Kalipatti and Criket Ball developed at UAS, Dharwad. Tree is vigorous, bearing round to slightly oblong fruits with high yield. The fruits are very sweet having a soft, granular and mellowing flesh with a TSS of 26° Brix. The colour of the pulp is light orange. The mean fruit weight is 150 g.

DHS-2: It is also a hybrid between Kalipatti and Cricket Ball. Tree is vigorous and bearing round fruits. It is a high yielder. The fruits are sweet with a TSS of 23° Brix having a light orange brown pulp, which is soft, granular and mellowing. The mean fruit weight is 180 g.

Chapter 20

Temperate Fruits

APPLE

Cultivated apple has been classified as *Pumila group*. Majority of the cultivated apples are diploids (2n=34) and few are triploids (2n=51). Delicious group of apples are very popular and occupy 50-70 per cent area in the states of Himachal Pradesh, Jammu and Kashmir, Uttar Pradesh and North-East hills.

Breeding Objectives

Apple is grown as a composite tree consisting of rootstock, scion and occasionally interstem. Thus genetic improvement must involve both rootstock and scion. The scion breeding objectives are to evolve varieties, red in colour with early maturity, high yield, superior dessert and storage quality and resistance to scab. Besides, a new wave of clonal rootstocks capable of surviving under wide range of environmental conditions, inducing precocity, enhancing productivity and fruit quality in scion are required to be bred.

Genetics Resources

Malus has 25 to 30 species and several sub-species, many of which are cultivated as ornamental trees for their profuse blossoms and attractive fruits. Many of the species intercross freely and semi self-incompatibility is common. Trees grown from collection of *Malus* are frequently inter-specific or inter-varietal hybrids. The cultivated apple is botanically *Malus domestica* Borkh.

The majority of cultivated apples are diploids (2n=34). There has been a belief that they are complex polyploids, being partly tetraploids and partly hexaploid with the basic number of x=7 which is common in Rosaceae. The hypothesis is based on the associations and behavior of chromosomes and six sets of three chromosomes.

So, they are functionally diploids. Among the cultivars, there are also triploids (2n=51). Triploids appear to be more common in cultivated apples, accounting for about 10 per cent of the commonly grown cultivars. Some triploid varieties are Baldwin, Gravenstein, Rhode Island Greening, Blenheim Orange and Mutsu. These are more vigorous and tend to have larger fruits but produce poor pollen and require diploids to pollinate them. These are useless as parents for breeding as they produce few seeds and give rise to weak seedlings.

Sterility and Incompatibility

Sterility and incompatibility are two main causes of unfruitfulness in apple. The generational sterility is caused by the failure of any of the processes concerned with the development of pollen, embryo sac, embryo and endosperm. This is common in triploids and some diploids. Gagnieu (1951) concluded that the segregation suggests a simple disomic inheritance of four different and possibly allelomorphic genes P^1 P^2 P^3 and P^4.

Sexual incompatibility which is due to the failure of the pollen, although functional, to grow down the style and bring about fertilization is widespread in the apple. Self-incompatibility is particularly common, although cases of cross incompatibility are also known.

Apomixis

Facultative apomixis is characteristic of a number of *Malus* species which are probably of hybrid origin but does not appear to occur among the cultivated apples. The apomictic species which have been investigated are polyploids. *Malus sikkimensis* (Hook) Koehne is a triploid, *M. coronaria* (L.) Mill., *M. hupenhensis* (Pamp.) Rehd., *M. lancefolia* Rehd. *M. platycarpa* Rehd., *M. toringoides* (Rehd.) Hugs are known in triploid and tetraploid forms. *M. sergenti* Rehd is known in diploid, triploid, tetraploid and pentaploid forms. Under normal circumstances, these species reproduce themselves freely by apomictic seeds but most of them can produce sexual hybrids if crossed with sexual diploids. Seedlings from these apomictic species are not necessarily identical and a certain amount of variation can be found. The importance of this character in *Malus* species is that seedlings of some are sufficiently uniform to enable their use as rootstocks which are virus free.

Breeding Methods and Achievements

Introduction and Selection

At Regional Fruit Research Station, Mashobra, spur varieties introduced through the National Bureau of Plant Genetic Resources (NBPGR), New Delhi during the eighties are under evaluation. These varieties include Red Spur Delicious, Golden Spur Delicious, Miller's Sturdeespur, Oregon Spur and Red Chief of which Red Spur Delicious has been found to be promising. In UP, cultivars Red Spur and Oregon Spur were introduced from Italy and are being multiplied for evaluation.

Colour Sports

Colour sports like Royal Red, Vance Delicious, Top Red, Skyline Supreme Red Delicious were introduced in HP. The cultivars Royal Red, Vance Delicious and Top Red and Skyline Supreme Red Delicious were found to be promising.

Early Varieties

Among the early varieties introduced at NBPGR Regional Station, Phagli, Shimla, EC 32221, EC 38683, Yandik-Ovskoe and Papisovka Canniaga are promising.

Low Chilling Varieties

Work at NBPGR Regional Station, Phagli, Shimla indicated that the cultivars Vered, Michal, Maayan, Shilomit, Hybrid-1 and Tropical Beauty were found to be promising for cultivation under mid and low hill conditions. In the mid hills of HP, the cultivars Tropical Beauty and Parlins Beauty were found to be the best in respect of yield and fruit quality.

Scab Resistant Varieties

Scab is a serious disease of apple and none of the commercial varieties is resistant to it. Some resistant varieties have been evolved in other countries, none of these compares favourably with the popular Delicious which is commercial sport. The scab resistant varieties Prima, Priscilla, Sir Prize, Jonafree, Liberty and Coop.12 introduced from USA are under evaluation at Regional Fruit Research Station, Mashobra and Bajaura in HP.

Hybridization and Selection of Parents

Most of the quality traits like size, shape, cropping, *etc.*, are under polygenic control. Thus, when two cultivars are crossed, there will be a continuous range of expression of these characters in the seedlings and will not segregate into discrete categories.

Williams (1959) calculated that the percentage of desirable seedlings that can be expected as the main product of an apple breeding programme for polygenically controlled characters is seldom more than 40 per cent and for every additional character, the figure rapidly decreases. Thus, for a programme in which the main objective is polygenically controlled mildew resistances, size of fruit, season of maturity, flavour and colour of skin, a reasonable estimate would be 40, 20, 20, 10 per cent respectively.

New Varieties

The modern breeding, objectives are breeding of varieties with high yield, superior dessert and storage quality, disease and pest resistance. Breeding work on apple has been in progress at Regional Fruit Research Station, Mashobra in Himachal Pradesh, Fruit Reseach Station, Shalimar in Kashmir and Horticultural Experiments and Training Centre, Chaubattia in Uttarakhand. The major objectives were better shelf-life, early maturity, high dessert quality and scab resistance.

Shelf-life and Dessert Quality

All the popular Delicious group of cultivars ripens at the same time and thus cause glut in the market. With a view to combine high dessert quality with good keeping quality, work was initiated in Kashmir in 1956. Two hybrids, Lal Ambri (Red Delicious x Ambri) and Agold (Ambri x Golden Delicious) were released. Work on similar lines was started in HP in 1960 (Chand, 1962). As a result three promising hybrids, namely, Ambred, Ambstarking, and Ambrich were selected. Subsequently, hybrid Ambroyal was also selected salient characters of these hybrids are enumerated below.

Ambred (Red Delicious x Ambri 157): Tree is tall, maturity in second week of September; fruits medium in size, conical, symmetrical, bright red stripes over barium yellow ground; dots obscure; skin medium in thickness, smooth and glossy; flesh whitish, crisp, firm aromatic and juicy, keeping quality is good up to three months in air cooled storage. It has low incidence of powdery mildew, sooty blotch and apple scab.

Ambstarking (Starking Delicious x Ambri 81): Tree is vigorous, tall and open, maturity in second week of September; fruits medium in size, round, conical symmetrical and uniform in shape, currant red streaks over chrome yellow ground; dots numerous and conspicuous; skin rough, smooth, flesh whitish firm, crisp, tough and juicy; keeping quality comparable with Starking Delicious. It is tolerant to apple scab.

Ambroyal (Starking Delicious x Ambri 84): Tree is semi-dwarf and spreading. Fruit maturity is in third week of September; fruits medium in size, conical in shape; skin thin, smooth, red streaks on yellow ground; flesh white, soft, sweet, juicy with good dessert quality. Storage quality is comparable with Starking Delicious.

Ambrich (Richard x Ambri 15): Tree is semi-dwarf, semi-spur type; spreading drooping fruit maturity in second week of September, fruit medium size, round, conical in shape, symmetrical sides equal and uniform, skin thick, smooth with chrysanthemum crimsonwash, flesh whitish, firm crisp, sub-acid aromatic and juicy with good dessert quality. Tree is moderately susceptible to powdery mildew and tolerant to apple scab.

Early and dessert quality: Work was started at Chaubattia in 1970 and two promising hybrids Chaubattia Princess and Chaubattia Anupam were evolved. Both these are from crosses of Red Delicious x Early Shanburry.

Chaubattia Princess ripens during last week of June to the 1st week of July. The tree is of medium vigour with upright growth habit. Fruits are medium in size, regular and conical in shape. Fruit skin is thin and smooth with deep red streaks on pale background. Flesh is creamy white, crisp in texture, firm juicy and very sweet. TSS is 14 °Brix and acidity 0.22 per cent. The fruit pressure at maturity is 14 to 15 lb/sq. inch. Keeping quality is quite good.

Scab resistance: During 1983, crosses Gala x 58553, Liberty x Delicious, Gala x 6356- 22 Gala x 6143-1, Freedom x Delicious, Gala x Prima and Freedom (open pollinated) were made at Mashobra and the hybrid seedlings are being evaluated.

Out Breeding and Backcrossing

Dominant single gene resistance in a *Malus* species can be transferred to the cultivated apple by a modified backcross procedure to avoid inbreeding. The method involves crossing the wild species with a large fruited cultivar. The resistance F_1s is heterozygous and the best ones are selected and backcrossed to a good cultivar and their progeny yields 50 per cent resistant seedlings. The best of these are again backcrossed to a good cultivar until all the good qualities of the cultivated apple are recovered and the resistance from the wild species retained. This avoids inbreeding by alternating different cultivars for the recurrent quality parent and eliminates loss of vigour and incompatibility problems.

Mutation Breeding

Work on induction and selection of desirable bud mutants was taken up at Horticulture Experiment and Training Centre, Chaubattia in 1973. As a result, four mutants with distinctly compact habit and better keeping quality of fruits were selected and are being evaluated under different agroclimatic conditions.

PEACH

Breeding Objectives

The main objective of peach (*Prunus persica*) improvement for low chilling areas would be to develop cultivars with low chilling requirement, tolerance to high summer temperature, maturity between 60 and 70 days after full bloom, firm flesh, freedom from loose fibre, attractive colour, non- browning of flesh, resistance to root-knot nematode, iron chlorosis and water logging. For processing peaches, firmness of flesh, freedom from loose fibre, attractive colour and non-browning of flesh are the important characters to be improved.

Breeding Methods and Achievements

Introduction and Selection

A large number of low chilling peach varieties, *e.g.,* Floridasun, Sun Red and Sun Gold and some other selections, Floridared and Floradabelle were introduced at the PAU, Ludhiana, during late sixties from Florida and California states in USA. Of these introductions, Floridasun, Floridared, Sun Red and 16-33 (Named-I-Shan, Punjab) became very popular. Of the later introductions from USA, TA 170, known as 'Partap', has been identified as early (7 days earlier than Flordasun). Its flesh is yellow, firm, with red colouration and better keeping quality. Another two introductions from Florida, Flordaprice and Earligrande, have been recommended for commercial cultivation for the plains of Punjab and adjoining areas. Flordaprice is early ripening, whereas Earligrande is an mid-season variety. Clonal selection 'Sharbati' is a chance seedling selected at Saharanpur.

Hybridization

Redhar is a cross between "Halehaven and Kalhaven bred at USA. Inter-specific hybridization has also been attempted in peaches especially in the development of rootstock resistant to nematodes. Nemaguard, a hybrid between *P. persica x P. davididasa* is a widely used root-knot nematode resistant rootstock, which is immune to *Meloidogyne incognita.* Planned hybridization work on peach was started in 1957 at Saharanpur. Peach Saharanpur Prabhat (Sharbati x Flordasun) was released. Fruits of this variety are attractive, sweet, maturing at least 4 days earlier than Flordasun.

PLUM

Breeding Objectives

In European plum (*Prunus domestica*), improvement for cold hardiness, productivity, large sized fruits, colour (red, purple or blue), free stone and dessert quality are important criteria. For Japanese plums (*P. salicina*), self fertile, late blooming plums, with high quality (particularly yellow skin) are important characteristics.

The main objectives of plum improvement programme for subtropical regions are to develop an early maturing cultivar with low chilling requirement, tolerant to high temperature and dwarfing rootstocks, tolerant to saline and stagnant soils, large fruited, free stone, juicy with proper TSS/acid ratio, suitable for processing and resistant/tolerant to insect, pests and diseases.

Breeding Methods and Achievements

Introduction

A large number of plum varieties have been introduced from different countries. Of these, Santa Rosa and Sutlej Purple are important commercial cultivars found suitable for midhills of North Western Himalayas. Other methods of breeding are not yet followed in this crop in India.

PEAR

Breeding Objectives

Pear, *Pyrus communis*, has a chromosome number of 2n=34. Breeding objectives are to develop dwarf scion and dwarf rootstocks tolerant to wet and saline soils and resistant to diseases like *Ganoderma* and root rot, free from low bud differentiation, alternate and shy-bearing of Baggugosha and Le-Conte and selection of superior clones of Patharnakh and Baggugosha.

Breeding Methods and Achievements

Important and popular cultivars such as Bartlett, Anjou, Kieffer are only introductions from Europe and are well acclimatized to the Northern and Southern Indian hills. A lot of variability, however, exists in soft pear plantations for yield, regular bearing, fruit size, shape, skin colour and fruit quality. An extensive survey of pear growing areas in Punjab and adjoining states taken up by the PAU, Ludhiana resulted in the identification of 19 superior strains of softpears. Of these, soft-fleshed selections 'Red Blush" 'Punjab Gold' and "Punjab Nectar" are promising. Red Blush recorded the highest yield (23.7 tones/ha) with good quality attributes.

Section-III

Breeding Behaviour of Vegetable Crops

Chapter 21

Tomato

Botanical name: *Solanum lycopersicum* (Mill.) Wettsd.

(syn. *Lycopersicon esculentum* Miller)

Family: Solanaceae

Chromosome number: 2n = 2x = 24

History of Tomato Breeding

Tomato is one of the most important vegetable crops grown throughout the world. The leading tomato growing countries in the world are the USA, several European countries, Japan and China. It is rich source of vitamins and minerals. Tomatoes are consumed as fresh and in processed form. The genus *Lycopersicon* differs from Solanum by the absence of spines in the pinnatified leaves and narrow anther tips, which exhibit longitudinal dehiscence. But again Peralata and Spooner gave the name *Solanum lycopersicum* L. in year 2000.

Botany

Tomato belongs to family Solanaceae and the genus *Lycopersicon*. At present nine species are recogniesed within this genus. Whereas, Muller (1940) divided into two sub-genera, *viz.*;

- **Eulycopersicon:** There is three red fruited edible species- *Lycopersicon esculentum,* (SP) *L. pimpinellifolium* (SP+CP) and *L. cheesmanii* (SP) which containing carotenoids and lycopene pigment.
- **Eriopersicon:** They are green fruited species with anthocyanin pigment. The species are *Lycopersicon peruvianum,* (SI) *L. glandulosum, L. hirsutam,*

(SF, SI) *L. chilense,* (SI) *L. parviflorum,* (SP) *L. chemielewskii* (CP) and *L. pinnellii* (SI).

(SP= Self-pollinated, CP= Cross pollinated, SF= Self-fertile and SI= Self-incompatible)

Inter-specific crosses between *Lycopersicon esculentum* and *L. pimpinellifolium* are easily made. Embryo abortion may occur between *Lycopersicon esculentum* x *Lycopersicon peruvianum.* However, this can be overcome through embryo rescue technique.

Tomato produces perfect hermaphrodite and pentamerous flowers. The anthesis starts around 6.00 am with maximum flower opening at 7.00-8.00 am. Anthesis is followed by dehiscence which takes place between 7-11 hours. Pollen grain remains viable for 2-4 days. Stigma becomes respective 15-20 hours before anthesis and until 4-7 days after anthesis, provided an optimum temperature of 8-25°C. From blooming to full ripening of fruit take about 45-50 days required. Natural cross-pollination varies from 2-5 per cent. In tropical climate cross-pollination has been reported up to 26 per cent however it is very rare. Emasculation is usually done one day prior to anthesis/flower opening. This stage the sepals have started to separate and the anthers and corolla are beginning to change from light to dark yellow. The stigma is fully receptive at this stage allowing for pollination even immediately after emasculation.

Germplasm Resources

Germplasm is the basic requirement to develop new varieties. It consists of wild species, landraces, old varieties, new varieties and advanced lines. The following are the germplasm resource centres at global level:

- NBPGR - National Bureau of Plant Genetic Resources, New Delhi, India.
- WVC- World Vegetable Centre (earlier AVRDC- Asian Vegetable Research and Development Centre), Taiwan, Republic of China.
- National Centre for Genetic Resource Preservation, Colorado, USA.
- North East Regional Plant Introduction Station, New York.
- USDA - United States Department of Agriculture, Washington DC, USA.
- TGRC - Tomato Genetic Resource Centre, California, USA.
- NIAS - National Institute of Agricultural Sciences, Japan.
- IVT - Institute of Horticulture and Plant Breeding, Netherlands.
- CGN - Centre for Genetic Resources, Netherlands.

Wild Species

Wild species are the valuable sources of resistance to biotic and abiotic stresses and are being used for the development of resistant varieties. Resistance and other desirable attributes reported in wild species are:

L. pimpenellifolium - Resistance to *Fusarium* wilt, Early blight, Leaf curl, Bacterial speck, low temperature tolerance and high ascorbic acid content.

L. cheesmanii- Resistance to salinity and high TSS.

L. peruvianum - Resistance to Leaf curl virus, TMV, Root knot nematode and rich source of ascorbic acid.

L. hirsutum - An excellent source of resistance to TMV, Bacterial canker and Bacterial speck.

L. hirsutum f. *glabratum* - Resistance to Leaf curl virus, Early blight, various insect- pests and tolerance to low temperature.

L. chilense- Tolerance to tomato Yellow leaf curl virus and resistant to drought.

L. chemieleskii- Rich source of TSS.

L. penneilii - Drought tolerance.

Breeding Objectives

- ✫ To develop early and high yielding varieties/hybrids having smooth, uniform fruits with good flavour, high vitamin A and C contents.
- ✫ Breeding varieties resistant to biotic and abiotic stresses.
- ✫ To develop varieties suitable for processing, *viz.*, deep red colour, titrable acidity as citric acid of about 1.75 per cent, pH 4.0-4.1, lower pro-vitamin A (β-carotene), high TSS and viscosity.
- ✫ To develop varieties suitable for transportation and have prolonged storage (thick fruit skin and pericarp).

Breeding Methods and Achievements

Various methods have been employed to develop varieties. However, application of breeding method depends upon the objective of the programme. Correlation studies among different characteristics reveal that number of fruits is positively correlated with yield. Total yield is positively correlated with early yield, fruit number/plant and fruit weight. Number of primary and secondary branches, plant height and number of fruits/cluster are positively correlated with yield. However, the maximum direct contributions towards total yield have been associated with total number of fruits/plant, and indirect effect with fruit size and plant height.

1. Introduction

Seeds of desired genotypes are introduced from one ecological area to another. The introduced genotypes are evaluated in different agro-climatic regions along with suitable check. Superior genotype is recommended for cultivation or utilized into the crop improvement programme. A large number of varieties have been introduced from abroad and became popular in different agro-climatic regions depending upon their adaptability.

Introduced Tomato Varieties

USA- Roma, Labonita, Sioux, Marvel, Money Maker, Chiku Grande, Best of All and Nematox.

Taiwan- VC 48-1.

Hawai- Sel-120.

Israel- Dwarf Money Maker.

2. Pureline Selection

Single plant selection followed by pureline selection till homozygosity is attained. A number of varieties have been bred by this method.

Tomato Varieties Developed by Pureline Selection

Indigenous- CO-1, CO-2, *etc.*

Exotic- Arka Vikash, Arka Saurabh, Labonita, Improved Meeruti, Pant Bahar, Pusa-120, Pant T-3, HS-110, *etc.*

3. Heterosis Breeding

Heterosis is the superiority of a hybrid over their parents in vegetation adaptiveness and productivity. Superiority is determined not by its quantitative value, but its significance in welfare of man. Hybrids have superiority over open-pollinated varieties for earliness uniformity, yield and resistance attributes. They may have better adaptability than open-pollinated varieties. Very earliness, large fruit size, uniformity and smoothness can be easily developed in hybrids in comparison to open-pollinated varieties. It is also possible to combine number of resistant genes in hybrid than pureline varieties. For development of hybrid varieties, male and female parents are selected. In case, if the parental lines are not pure, it may be achieved by 3-4 generation of selfing. After selection of parent, general combining ability (gca) of lines and specific combining ability of crosses are studied. Practically the superiority of a hybrid is worked out by the *per se* performance of mean value. High gca lines are utilized to exploit additive gene action by selecting after hybridization.

Hybrid Seed Production

(i) Hand Emasculation and Hand Pollination

By and large, hybrid seeds are produced by hand emasculation and hand pollination. First, the female parent is selected. Female parent should have high seed yield with good germinating ability and resistance to diseases, especially to seed-borne diseases. Flower drop after fertilization in seed parent should be prevented. Then male parents are chosen which complement those characteristics that are not inherited in the female parent. It is a costly method of seed production, and for one hectare of hybrid tomato seed production, requires 5 skilled pollinators for 30 days or equivalent to 12 thousands hours of labour. Even the use of male sterile lines in commercial production reduces the labour need by only 50 per cent, because pollination has to be done by hand. Seed parents (female) and pollen parents (male) are grown in the ratio of 10-12:1. Pollen parents are generally sown one to two weeks earlier than the female parents to provide sufficient pollen, when the seed parents are ready for crossing.

Emasculation: Those flowers whose sepals are opened at 30-40° angle with flower axis are taken for emasculation. Practically the flowers of seed parents are taken for emasculation 12-15hr. before opening. Anthers are removed by forceps or needles. Generally, emasculation is done during evening.

Pollen collection and pollination: Pollen collecting apparatus (vibrating pollen collector) is used which collects pollen from 1000 flowers in one hour. In a small-scale operation, pollen is collected with needle and forceps. Pollen is mostly collected in afternoon. At room temperature and 70-80 per cent relative humidity pollen can be stored for 5 to 6 days. Generally, freshly collected pollens are used. A needle, brush or forceps is generally used for pollination, but small blackened dental spoons and J shaped glass tubes are most useful for direct transfer of pollen to the stigma. Pollination instruments are sterilized in ethanol before and after use. Pollen of one male flower is sufficient for the pollination of 12-18 flowers. Temperature, humidity and wind velocity affect pollination. High wind velocity reduces the pollination efficiency. The stigma must be receptive at pollination. Petals serve as an indicator of stigma receptivity. When the petals of the emasculated flowers start to turn bright yellow, the stigma is ready for pollination. Soon after pollination, the pollinated flowers are tagged. Two or three sepals are clipped to serve as a marker for hand pollinated fruits. All flowers that escaped emasculation are removed as soon as possible to avoid any contamination.

(ii) By Use of Male Sterile Line

Genetic male sterility is found in tomato. In this case, pollen grains are sterile. Male sterile plants have small flowers, poorly developed anthers, poor or no anther dehiscence. Such lines are maintained by crossing fertile plants (Ms ms) with male sterile parent (ms ms), which produces 30 per cent fertile and 50 per cent sterile plants. The sterile plants are used for production of hybrids and fertile plants are removed. The flowers of sterile plant are pollinated by the pollen of male parent to produce hybrid seed. The potential sterile line known is ms10.

(iii) By Use of Functional Male Sterile Lines

Hybrid seed may be produced by use of functional male sterile lines with exerted style 'ps' which have shortened stamens, non-opening or non-dehiscent anther type "ps2". Such lines can be produced by spraying of 1000 ppm GA_3, 6 days before blooming. The flowers are pollinated by pollen of male parent to produce hybrid.

Harvesting and Seed Extraction

Only fully red ripe fruits that are hand pollinated are harvested and seeds are extracted by fermentation method.

4. Hybridization

For hybridization, male and female parents are selected on the basis of general combining ability. Crosses are made and F_1 seed is harvested. F_1 and subsequent generation are raised. Selection of desirable plants is done following pedigree method or single seed descent method in segregating population. In pedigree method, single plant selection is done in F_2 and is continued in successive generations till purelines are obtained (generally up to F_6). In early generations, the selection emphasis is on highly heritable characters like growth habit, resistance to diseases, fruit characteristics, *etc.*, whereas the characters with low heritability like fruit yield and quality are taken into account in later generations. In single seed descent method (also referred to as modified pedigree method). F_2 population is planted and one seed is obtained from each plant to grow in next generation till F_6 in which individual superior plants are selected for individual plant progeny planting and selection of superior progeny rows in the next generation.

In single seed descent method, generations can be advanced in the off-seasons and this allows maintenance of broad genetic base in advanced generation. In recent past most of the tomato varieties have been bred by hybridization followed by pedigree selection.

Tomato Varieties Developed through Hybridization

Variety	*Parent Involved*
Pusa Early Dwarf	Improved Meeruti x Red Cloud
Pusa Ruby	Sioux x Improved Meeruti
HS-101	Sel.- 2-3 x Exotic cultivar
HS-201	Sel. - 12 x Pusa Early Dwarf
Hisar Arun (Sel.7)	Pusa Early Dwarf x K2
Punjab Chhuhara	Punjab Tropic x EC 55055
Hisar Lalima (Sel. 18)	Pusa Early Dwarf x HS-101
Hisar Lalit	Resistant Bangalore x HS-101
Sweet-72	Pusa Red Plum x Sioux
Pusa Gaurav	Glamour (exotic) x Watch (exotic)
Panjab Kesari	Punjab Tropic x EC 55055
Marglobe	Marvel x Globe
Keck-Ruth Ageti	Kachmethi x Rutgers
Pusa Red Plum	*L exculentum* x *L pimpinellifolium*
Sel.- 2	(HS-101 x Punjab Tropic) x (H-14 x Punjab Tropic)
Pusa Sheetal	Balkan (exotic, Bulgaria) x Jemnornosnej (exotic. Russia)

5. Mutation Breeding

Tomato genotypes are highly responsive to mutagenic treatments. After treating the seed with mutagens, it is sown for raising M_0 population. Seeds are extracted and subsequent generations, *viz.,* M_1, M_2, M_3, are grown. Selection of desirable plants is done by following pedigree method or single seed descent method in segregating populations.

Varieties Developed through Mutation Breeding

Varieties	*Mutant Type*
S-12	X-ray mutant of Sioux
Maruthan (CO-3)	EMS mutant of CO-1
PKM-12	Mutant of Annagi
Pusa Lal Meeruti	Gamma ray mutant of Meeruti

5. Breeding for Quality

Tomato fruit quality for fresh consumption is determined by appearance (colour, size shape, freedom from physiological disorders and decay), firmness, texture, dry matter and organoleptic nutraceutical (health benefit) properties. Texture is greatly influenced by fruit internal structure (pericarp: locules ratio), cuticle thickness and flesh firmness. The organoleptic quality of tomato is mainly attributed to its aroma, volatiles, sugar, acid, vitamin, carotenoid, and flavonoid content. Post-harvest durability and fruit safety are also very important quality criteria for product distribution and marketing. The major quality components are; TSS, sugar, acidity, lycopene and carotenes. Among soluble solids, glucose and fructose are the major components whereas protein, cellulose, pectic substances and hemicelluloses are the main component of insoluble solids. However, sugar/ acid ratio decides flavour of the fruit. Combinations of at least 16 aroma compounds together give tomato its unique flavour. High TSS is desired in processed tomatoes. Processed tomatoes should have total solids over 6 per cent, pH below 4.3, solids/ acid ratio of 15, a sugar/acid ratio of 8.5 and high viscosity. *L. chemielewskii* (10 per cent TSS) and *L. cheesmannii* (15 per cent TSS) are good donor parent to increasing TSS. In commercial cultivars TSS ranged from 4 per cent to 6 per cent. In processed tomatoes, pH value should be 4.0-4.1 and titrable acidity 1.75 per cent as citric acid. Concentrated flowering, uniform fruit set and ripening and resistant to fruit cracking are the important selection criteria for processing varieties. Ascorbic acid content is also very important factor as for as quality is concern. It ranges from 10 to 20 mg/100 g of fresh weight of fruit. *L. peruvianum* and *L. pimpinellifolium* are good sources of ascorbic acid. Elongated fruits are less damaged in transportation, thus they are more suitable for long-distance transportation and storage.

6. Selection of Tomato Cultivars

The following criteria are considered in selection of tomato cultivar:

- ✰ Resistance to diseases and physiological disorders.
- ✰ Yield (early as well as high total yield).
- ✰ Short-term and long-term plant vigour.
- ✰ Growth rate.
- ✰ Fruit colour.
- ✰ Fruit size and shape.
- ✰ Fruit flavour and
- ✰ Shelf-life.

The final choice is often a compromise between market factors, agronomic criteria and production constraints.

Chapter 22

Brinjal

Botanical name: *Solanum melongena* L.

Family: Solanaceae

Chromosome number: 2n = 2x = 24

History of Brinjal Breeding

Brinjal once deemed to be poor man's vegetable, has occupied a prominence place in routine diet of all classes of society. Brinjal is a typical native vegetable grown almost throughout the year. It has a regional preference, and therefore, one has to breed varieties as per regional preference. Before start of a brinjal breeding programme, survey of major brinjal growing areas and their markets is necessary.

Botany

Brinjal is grown as herbaceous annual while it is perennial in the tropics. Inflorescence is often solitary but sometimes it constitutes a cluster of 2-5 flowers. Solitary or clustering nature of inflorescence is a varietal character. Flower is complete actinomorphic and hermaphrodite. Calyx is five lobed, gamosepalous and persistant. It forms cup-like structure at the base. Corolla is five lobed gamopetalous with margins of lobes incurved. There are five stamens which are free and inserted at the throat of corolla. Anthers are cone shaped, free and with apical dehiscence. Ovary is hypogynous, bicarpellary, syncarpous and with basal placentation.

Heterostyly is a common feature. Fruit setting flowers consist of long and medium styled flowers. Fruit setting long styled flowers normally varies from 70-80 per cent and that in medium styled flowers from 12-55 per cent. The non-fruit setting flowers consist of short styled flowers in which androecium are fertile but stigma is smaller with under developed papillae. Brinjal is considered as self-pollinated crops but the extent of cross-pollination has been reported as high as 29 per cent hence it is classified as often cross-pollinated or facultative cross-pollinator.

Flowers generally emerge 40-45 days after transplanting. Anthesis occurs at about 6.00-8.00 am in August-September and usually between 9.30-11.15 am during winter (December-January). Stigma receptivity is highest during anthesis, *i.e.*, flower opening. Anthers usually dehisce 15 to 20 minutes after the flower have opened. The receptivity of the stigma can be observed from its plump and shiny appearance which gradually becomes brown with the loss of receptivity. The period of effective receptivity ranges from a day prior to flower opening until about four days after opening. Pollen usually remains viable for a day during summer and 2-3 days in winter under field conditions.

For emasculation a healthy long or medium styled well developed bud from the central portion of the plant is selected. The bud is opened gently with the help of fine pointed forceps one or two days before the opening of the bud and all the five anthers are carefully removed. For pollination freshly dehiscing anthers are picked up and are slit vertically with fine needle to get sufficient pollen at the tip of the needle. Pollen are applied on the stigma of the emasculated flower bud. It is labeled and covered with small pollination bag.

Germplasm Resources

In India, the National Bureau of Plant Genetic Resources (NBPGR), New Delhi is actively involved in building up of germplasm with wide variability through exploration and introductions its proper characterization and evaluation and subsequent identification of suitable doners types for utilization by users in crop improvement programme. Besides, IIVR, Varanasi, IIHR, Bangalore are also maintaining the germplasm of brinjal. Globally Asian Vegetable Research and Development Centre, Taiwan and Chinese Crop Germplasm Information System (CGRIS) are maintaining the germplasm.

Breeding Objectives

The brinjal varieties are bred to shake up the following objective:

- Earliness.
- High yield.
- Fruit shape, size, colour as per consumer preference.
- Low proportion of seeds.
- Soft flesh.
- Lower content of alkaloids (α- solasonic and α- solamargine).
- Upright sturdy plant free from lodging.
- Resistance to biotic and abiotic stresses.

Breeding Methods and Achievements

1. Pureline Selection

Pureline selection generally practiced to make improvement in land race/ heterogeneous materials. In this method, seeds are sown and seedlings are transplanted at wider spacing. Individual plants are examined at full fruiting stage and superior individual plants based on earliness, fruit shape, size, colour, yield (visual estimate) plant type resistance to diseases/pests, *etc.* are selected and harvested separately. Next year individual plant progenies are planted in rows (3 rows of 10m length) and uniform progenies with desirable traits are selected and seeds within a selected progeny row are bulked and a new line is thus constituted for further trials. In case some individual plant progenies show segregation individual plant selection may be repeated.

Varieties developed through pureline selection: Pusa Purple Long, Pusa Purple Cluster, Pusa Purple Round, Pant Samrat, Arka Kusumkar, Arka Sheel, Punjab Chamkila, Punjab Neelam, Punjab Bahar, Krishnanagar Green, *etc.*

2. Hybridization

Desirable male and female parents are selected on the basis of general combining ability. Crosses are made between two parents. Mature crossed fruit are harvested and seed extracted. Large F_2 population is grown from which individual plants are selected on the basis of phenotypic characters. Individual single plant selection process or single seed descent methods can be used to advanced the generation up to F_6 or F_7. The selection is early generation is based on earliness, fruit size and yield. However quality, disease and pest incidence are also taken into consideration. After attaining high degree of uniformity from F_6-F_7, the superior families are selected and evaluated along with suitable check (standard variety) up to F_{10} generation. A few best selected lines are then tested at several locations alongs with suitable check. Best performing lines is identified for release on the basis of three consecutive year performance in multi-location trials.

List of Varieties Developed through Hybridization

Varieties	*Parents Involved*
Pusa Kranti	(Pusa Purple Long x Hyderpur) x Wynad Giant
Hishar Shyamal	Aushey x BR-112
Hishar Jamuni	Aushey x R-34
Pant Rituraj	T-3 x Pusa Purple Cluster
Pusa Anupam	Pusa Kranti x Pusa Purple Cluster
Punjab Barsati	Pusa Purple Cluster x H-4
Sadabahar Baigan	Japanese Long x R-34
Pusa Uttam	GR x Pant Rituraj
Pusa Bindu	GR x Pant Rituraj
Pusa Upkar	GR x PB 91-1

3. Heterosis Breeding

Nagia and Kida were the first researchers who observed hybrid vigour in brinjal during 1926. Now-a-days, evaluation of heterosis has become a potential tool for improvement of brinjal in respect to higher productivity, resistance, uniformity and adoption. Generally, heterosis for yield increases with increase in number and size of fruits. The magnitude of heterosis varies with crosses environment and parents. High estimates of heterosis have been observed for yield number of fruits, fruit weight and prolonged fruiting period. Heterosis for yield varies from 80-240 per cent as reported by different research workers. Considerable heterosis has also been reported for pollen grain size, leaf size, number of leaf, plant spread, stem thickness, number of branches, earliness in fruit production, fruit set, fruits/ plant, fruit length, equatorial diameter of fruit, average fruit size, mean fruit weight, vitamin C content, sugar content and total soluble solid content.

Many F_1 hybrids with small fruited like, MHB 10, MHB 39, ABH 1, ABH 2, Hybrid 2, Sumex 9, Sumex 19 and Aruna were evaluated under AICRP (VC) and are available in the market. In round fruited brinjal, the hybrids NDB Hybrid 1, Nembakar, Pusa Hybrid-6, and in long fruited brinjal, Pusa Hybrid-5, ARBH 2, Pant Hybrid 1, HOE 1404 and ARU I have been developed.

Hybrid Seed Production

There are four methods of hybrid seed production. However, hand emasculation and pollination is commercially adopted.

(i) **Hand emasculation and hand pollination:** Male and female parents are selected as per breeding objectives. The flower buds of the female parent which are intended to open on the next morning should be selected. They are emasculated (removal of pollen) and bagged. Emasculated buds are pollinated in the morning with pollen grains of the male parent and again bagged to prevent cross-pollination.

(ii) **Pollination without emasculation:** Pollination without emasculation at bud formation stage may be done. It may give up to 97 per cent hybrid seed.

(iii) **Using male sterile lines:** Male sterile line can be produced for the production of hybrid seed. It is controlled by a single recessive gene. Male sterility can be induced by spraying the flower bud with 10 ppm 2,4-D without causing female sterility. Hybrid seed is produced by hand pollination.

(iv) **Using exerted lines:** Exerted lines such as, BR-112 and PBR-91-2 can be used for the production of hybrid seed.

4. Inter-specific Hybridization

The related species of *S. melongena* have been used in breeding for disease resistance. *Solanum melongena* is crossable with *S. incanum* (syn. *S. coagulans*),

S. torvum and *S. macrocarpon* and *S. leneatum* when used as a female parent. Among these wild species, *S. incanum* used either as male or female parent is easily crossable with *S. melongena* indicating a close relationship. On the other hand, it is difficult to cross *S. melongena* with other wild species *viz., S. indicum, S. sisymbriifolium, S. viarum* and *S. xanthocarpum*, which carry resistant genes to insect pest and diseases.

5. Breeding for Quality

Colour, shape and softness are the important physical factors that govern quality. Besides, biochemical constituents such as anthocyanin, total phenol, polyphenol oxidase activity and glycoalkaloids content are also important for fruit quality. Glyco-alkaloids; α-solasonin and α-solamargin are linked with flavour quality. The alkaloid content should be low. For 'bharta' (baked or boiled and mashed) preparation soft brinjal with less seeds are preferred. In brinjal, total soluble sugars varied from 1.0 mg to 3.53 mg/100 g., free reducing sugars 0.38 mg to 1.73 mg/100 g., anthocyanin 0.16 to.04 at 550 nm in 10 per cent (w/v) aqueous extracts, total phenols 0.44g to 0.138 g/100 g. and glycoalkaloid 6.25 to 20.50 mg/100 g. There should be low level of peroxidase, polyphenol oxidase and orthodihydroxyphenolic compounds activities. It has been observed that additive gene action is important for total chlorophyll, chlorophyll a and b, whereas both additive and non-additive gene actions are involved in the inheritance of anthocyanin. The narrow sense heritability estimates are medium for these characters.

Chapter 23

Chilli

Botanical name: *Capsicum annuum*

Family: Solanaceae

Chromosome number: 2n = 2x = 24

History of Chilli Breeding

Chilli is most important vegetable crops from several view points. Chilli are constituents of many foods, add flavour, colour, vitamin C and pungency to world food industries. Chilli is the important crop of USA where the five major producing states are Florida, California, Texas, New Jersey and North Carolina. The cultivars of *Capsicum annuum* include both hot pepper syn. chili or chilli (pungent fruits) and sweet pepper syn. capsicum, (non-pungent fruits). In India also, *C. annuum* is commonly cultivated for commercial purpose. The other four domesticated species (*C. frutescence* L., *C. chinense* Jacq., *C. baccatum* L. and *C. pubescens* Ruiz and Pavon) are also grown in different regions, although very limited scale. India is the largest producer, consumer and exporter of chilli among other major producers in the world, and is a top in terms of international trade. Sweet, non-pungent peppers are widely used at the immature or mature stage as a vegetable.

The five major cultivated species are derived from different ancestral stocks found in three distinct centres of origin as given below:

C. annuum : Maxico

C. chinense : Amazonia

C. frutescence : Amazonia

C. baccatum : Peru and Bolivia

C. pubescens : Peru and Bolivia

Wild forms of all the five species except *C. pubescens* exist. Peppers are found growing wild from the southern edge of the USA to central Argentina. The species commonly grown in Mexico, the USA, Europe and Asia is *C. annuum*.

Chilli is cultivated for various market types and exhibits wide range of genetic and morphological diversity in terms of fruit size, shape and their consumption patterns. The chilli and capsicum are important crops of India. Productivity levels of both the crops are low. The regions may be poor genetic stock, incidence of large number of diseases and insect pests, non-availability of high yielding and disease and pest resistance open pollinated and hybrid varieties and lack of scientific package of agronomical practices. The various factors responsible for low yield may be tackled through a systematic breeding programme.

Botany

Chilli are short lived perennial herb, but grown as an annual crop at commercial scale. Pepper is a dicotyledonous plant of epigeal germination. Fruit are non-dehiscent, many seeded berry, hollow, pendulous, variable in shape, size, colour and degree of pungency. Although normally considered a self-pollinated crop, natural cross-pollination up to 90 per cent many occur depending on bee activity and heterostyly. Hence chilli should be considered as an often cross-pollinated crop.

The flowers of wild *Capsicum* species are pentamerous. However large fruited cultivar may have 5-7 corolla lobes. The stamens are 5 and alternate with the petals. Crosses can be made any time during the day but morning hours are preferably. Flowers are emasculated in bud stage. Pollen is transferred to the stigma either from mature un-dehisced anther by scooping it out through the lateral sutures with the needle or by touching a freshly dehisced anther to the stigma with the forceps.

Germplasm Resources

Germplasm is the basic requirement to develop new varieties. It consists of wild species, landraces, old varieties, new varieties and advanced lines. The following are the germplasm resource centres at global level:

- ✰ NBPGR - National Bureau of Plant Genetic Resources, New Delhi, India.
- ✰ IIVR- Indian Institute of Vegetable Research, Varanasi.
- ✰ IIHR- Indian Institute of Horticulture Research, Bangalore.
- ✰ ICARNEH- Indian Council of Agricultural Research complex for North East Hill Region, Shillong.
- ✰ WVC- World Vegetable Centre (earlier AVRDC- Asian Vegetable Research and Development Centre), Taiwan, Republic of China.
- ✰ National Centre for Genetic Resource Preservation, Colorado, USA.
- ✰ USDA - United States Department of Agriculture, Washington DC, USA.
- ✰ NIAS - National Institute of Agricultural Sciences, Japan.
- ✰ IVT - Institute of Horticulture and Plant Breeding, Netherlands.

Breeding Objectives

The objectives of chilli and capsicum improvement may be described on the basis of the horticultural groups of *Capsicum* spp.,

- ✰ Earliness.
- ✰ Desirable fruit shape and size (oblate or round fruit in bell pepper and long fruit in chilli).
- ✰ Superior fruit quality (pleasing flavour, high sugar/acid ratio, high pigment content and vitamin C in bell pepper and high capsaicin, a fat soluble, flavourless, odourless and colourless compound).
- ✰ High oleoresin in chilli.
- ✰ Resistance to biotic and abiotic stresses.

Breeding Methods and Achievements

Breeding Strategies

Presently pepper breeding programmes have relied on a relatively narrow genetic base within cultivars or various market types, although huge morphological diversity is present within (intra-specific) and between (inter-specific) species. This is because of (i) traditional market demand for specific fruits size and shape (ii) the use of pureline or backcross breeding within open pollinated commercial varieties and development of inbreds from the commercial hybrid and their utilization of recycled parental lines. Specifically, introduction, mass selection, recurrent selection, pedigree method, heterosis breeding and backcross breeding are being mostly utilized to develop varieties with improved yield and resistance to certain diseases.

Various methods have been employed to develop varieties. However, application of breeding method depends upon the objective of the programme. Correlation studies among different characteristics reveal that number of fruits is positively correlated with yield. Total yield is positively correlated with early yield, fruit number/plant and fruit weight. Number of primary and secondary branches, plant height and number of fruits/cluster are positively correlated with yield. However, the maximum direct contributions towards total yield have been associated with total number of fruits/plant, and indirect effect with fruit size and plant height.

1. Pureline Selection

This method is applicable to landraces/local cultivars being grown by farmers. In this method, initial seed stock is spaced planted and superior plants are selected and harvested separately. Next year individual plant progenies are grown and progeny showing superior performance and devoid of genetic variability is bulk harvested and evaluated further with check cultivars in replicated trial. Several chilli varieties in India have been developed by this method. These varieties are G-1, G-2, G-3, G-4, NP-46A, K-1, CO-1, Musalwadi, Sindhur, Patna Red, *etc.*

2. Pedigree Method

This method involves selection of superior plants in the segregating generation following hybridization between superior cultivars along with maintenance of pedigree record. Selection of superior parental cultivars is crucial step in this method. Such chilli cultivars are Andhra Jyoti, Pusa Jwala, Pusa Sadabahar, X-235, K-2, Pant C-1, Punjab Lal, *etc.* Therefore in pedigree method, as far as possible both the parental cultivars should be promising and superior types.

3. Backcross Method

This is normally used to transfer single/few genes from primitive cultivars/ wild forms to leading cultivars. In some cases even BC_2 families may be routed through pedigree method of breeding (modified backcross) instead of following routine backcrossing programme which needs 5-6 backcrosses with the recurrent parent.

4. Heterosis Breeding

F_1 hybrids of chilli are popular in the USA and Europe and are gaining popularity in India after the initiation of research and seed production work in vegetables by a large number of private sector seed companies. The first hybrid crop in India was Bharat developed by Indo-American Hybrid Seeds, Bangalore (1973) followed by marketing of large number of hybrid cultivars by several others seed companies. The greater success of hybrid cultivars in chilli could be attributed to:

- ☆ Sufficiently large flowers, easy emasculation and pollen in abundance.
- ☆ Large number of crossed seeds/fruit.
- ☆ Heterosis for yield.
- ☆ Easy development of dominant genes conferring resistance to diseases.
- ☆ High yield potential of hybrids under better management and inputs.

5. Mutation Breeding

Chilli genotypes are highly responsive to mutagenic treatments. After treating the seed with mutagens, it is sown for raising M_0 population. Seeds are extracted and subsequent generations, *viz.*, M_1, M_2, M_3, are grown. Selection of desirable plants is done by following pedigree method or single seed descent method in segregating populations.

Varieties Developed through Mutation Breeding

Variety	*Mutant Type*
California Wonder, K-1	X-ray mutagen
California Wonder, Sathur Samba	EMS mutagen
Zlaten Medal, NP46A, Albena	Gamma ray mutagen

Chapter 24

Cauliflower

Botanical name: *Brassica oleracea* var. *botrytis*

Family: Brassicaceae

Chromosome number: 2n = 2x = 18

History of Cauliflower Breeding

Cauliflower has been in cultivation in India since last 170 years. It was introduced from England in 1822 by Dr. Jemson, Incharge of Company Bagh Saharanpur, U.P. In India cauliflower is grown in two distinct environments, *viz.*, Indian cauliflower which is grown in the temperature range of 20-35°C, European types/Snowball group which is grown in the temperature range of 10-25°C. Harvesting is done from August or early September to March in northern plains and from March-November in hills. Early varieties are heterogeneous and there is lack of stability. Compared to other cole crops, hybrid varieties in cauliflower are few in number. Besides, there is wide gap in yield, and crop also suffers from various biotic and abiotic stresses. All these factors require urgent attention of cauliflower breeders for improvement.

Major Types of Cauliflower

1. Indian Cauliflowers

Indian cauliflowers are characteristically different from the types grown in Europe as they are tolerant to high temperature and humid condition and are perhaps the earliest maturing types known. Indian cauliflowers are dwarf selections of Erfurt or Snowball types.

2. European Cauliflowers

Systematic and extensive cultivation of cauliflower was first started in Italy where the originals were developed. These original Italian types were taken to France, England, Germany and Netherland where some important local types were developed from the, *e.g.,* the Northerns in Yarkshire and Darbyshire, the Cornish in Cornwall, the Angers and Roscoff in Brittany and the Erfurt or its allied Snowball in Germany and Netherland.

Differences between Indian and European Cauliflowers

Indian Cauliflower	*European Cauliflower*
Tolerant to heat	Susceptible to heat
Curd formation at and above 20° C	Curd formation at and above 5- 20° C
Annual	Biennial
Yellow, creamy and comparatively loose curds often with strong flavour	Snow white curds with very mild or no flavour
More variable	Less variable
More self-incompatible	Less self-incompatible
Small juvenile phase	Long juvenile phase
No need of vernalization for flowering but needs cold treatment at 10-16° C for about 6 weeks	Require low temperature below 10° C for about 8 weeks for flowering

Botany

The stem of the vegetative cauliflower plant is rather short, and it thickens to about the same extent as that of cabbage. The leaves are large generally oblong the younger ones being nearly always sessile. In contrast to situation in other cole crops, buds do not usually arise in the leaf axils. During transition to the generative phase which in contrast to the situation in most other cole crops, is accomplished in many types of cauliflower in the first year the peduncles in the axils of the bracts formed by the main growing point branch repeatedly so that branches even of the fifth order can arise. The young curd is at first entirely covered by the foliage on becoming visible it already has a diameter of over 5 cm. After sometimes the flower stems elongate and a number of the apices develop into normal flowers. There are 4 sepals, 4 petals, 6 stamens and 2 carpels. The carpels form a superior ovary with false septum and two rows of campylotropous ovules. The androecium is tetradynamous, *i.e.,* there are two short and four long stamems. The bright yellow petals become 15-25 mm long and about 10 mm wide. The sepals are erect.

The buds open under the pressure of the rapidly growing petals. This process starts in the afternoon and usually the flowers become fully expanded during the following morning. The anthers open a few hours, later the flowers being slightly protogynous. The flowers are pollinated by insects, particularly bees, which collect pollen and nector. The nector is secreted by two nectarines situated between the basis of the short stamens and the ovary. Situated outside the basis of the pairs of long stamens are also two nectarines, but these are not active.

Breeding Objectives

- ✰ To develop high yielding varieties.
- ✰ To develop self incompatible but cross compatible inbred lines and cytoplasmic male sterile lines as better combining parents for production of hybrids.
- ✰ To standardize techniques for hybrid seed production.
- ✰ To develop varieties and hybrids with self blanching type having better curd qualities.
- ✰ To develop varieties resistant to (i) biotic stresses (a) diseases like black rot, sclerotinia rot, alternaria blight, downy mildew and erwinia rot (b) insects spodoptera, semilooper, diamond back moth and aphids, (ii) abiotic stresses high temperature and rainfall stress (curding in August/ September).
- ✰ To develop suitable varieties for curd formation in summer and rainy season in the hills.

Germplasm Resources

Crucifer Genetics Cooperative (CrGC) is active in acquiring, maintaining and distributing seed stocks and pollen of Crucifer vegetables. An informative research book is published once every three years by CrGC. A comprehensive base collection of the *Brassica oleracea* including cauliflower is maintained at Horticulture Research International, Wellesbourne, and is recognized by the International Plant Genetic Resource Institute, Rome. Centre for Genetic Resources (CGN) the Netherlands and USDA/ARS National Plant Germplasm System (NPGS) also maintain germplasm of cole crops. In India, NBPGR is the nodal agency for germplasm collection and conservation. In addition, IARI Regional Station, Katrain and Dr. Y.S. Parmar University of Horticulture and Forestry, Solan are also maintaining cauliflower germplasm.

The major centers which are actively engaged in crop improvement work of cauliflower are Division of Vegetable Crops, IARI, New Delhi and GBPUA and T, Pantnagar. The work for the improvement of temperate types is mainly done at IARI Regional Station, Katrain (Kullu Valley) and Dr. Y.S. Parmar University of Horticulture and Forestry, Solan (H.P.).

Pollination Techniques

Selfing: The self compatible varieties of cauliflower can be selfed simply by bagging the flower stalk. Selfing is also done by caging some plants with flies in cages or by isolation planting of lines having decreased level of self-incompatibility.

Crossing: The flowers may be emasculated by removing 6 stamens using a pair of forceps. In self compatible cauliflowers (European types), the stamens are removed before the opening of the buds as the flowers are already fertile in the bud stage, crossing can be done at the same time. In self incompatible types, emasculation may be emitted. When pollination cages are available, crosses between self-incompatible types can be made by insects such as honey bees, bumble bees and flies.

Breeding Methods and Achievements

1. Introduction

Several cultivars of cauliflower have been introduced from foreign countries and acclimatized in our conditions for commercial cultivation. Some of these cultivars are improved Japanese and D-96 from Israel, Giant Snowball and Snowball-16 from Holland and Dania Kalimpong from Denmark.

Population Improvement

Initially mass selection was widely used for improvement of cauliflower in India. However, it was found slow and less effective for the improvement of polygenic attributes. Consequently, modifications like mass pedigree and family selection methods were adopted. These were based on progeny testing. These methods are also applicable to open pollinated heterogeneous stocks having sufficient variability due to self-incompatibility. It order to make the best use of additive gene action, recurrent selection is being followed for the improvement of a number of quantitative traits. This method is effectively used to improve the base population when a plateau is reached for yield and other economic attributes. Based on the results of heritability and genetic advance, it has been found that recurrent selection is useful to improve compactness, diameter, depth and weight of cauliflower curd and improvement in yield from 18-47 per cent over the original material.

2. Hybridization

This is applicable to European types which have self-compatibility. Two self-compatible lines are crossed and the hybrid progeny is subjected to simple pedigree/bulk/backcross method of breeding. The cultivars Pusa Shubhra, Pusa Snowball-1 and Pusa Snowball K-25 were developed through hybridization. Pusa Shubhra was developed from a triple cross (MGS2-3 x 15-1-1) x D-96 and Pusa Snowball-1 was developed from a cross between EC 12013 x EC 12012 by making selections in F_2 and subsequent generations. Likewise, Pusa Snowball K-25 is a cross of EC-103576 a heading Broccoli and Pusa Snowball K-1.

3. Heterosis Breeding

This is based on the principles of developing inbred through bud stage selfing and production of F_1 seed by crossing/inter planting two self-incompatible but cross compatible inbred. Initially, except some Japanese varieties, none was commercially successful, and it was only after 1985 that F_1 hybrid cauliflowers became important. There is lack of F_1 cultivars in cauliflower in contrast to other *Brassica* vegetables due to following factors.

- Selfing of parental plants or sib-mating within the parental lines giving rise to off types after planting of commercial hybrid seed.
- Less effective self-incompatibility in cauliflower than in other brassicas.
- Non-synchronous flowering of male and female parents leading to increased proportion of sibs.
- The inflorescence of cauliflower tends to be cymose than recemose as in other brassicas, which results in a flush of flower production over a short length of time, leading to greater tendency for non-synchrony of flowering of the parents of the hybrid.
- Minor heterosis for curd size in contrast to substantial heterosis in other brassicas (Cabbage, Chinese cabbage, Brussels sprouts, *etc.*) which are grown primarily for the vegetative leafy tissues.

Heterosis breeding utilizes mainly dominant variance. Superiority in F_1 hybrids of cauliflower has been observed for earliness, curd weight, yield, resistance to diseases, hardiness and uniformity. Japanese seed industries have developed a number of F_1 hybrids in tropical types, improved both in quality and yield. F_1 hybrids have been developed utilizing the strong self-incompatibility system. Some of these like Snow Queen, Snow King, Snow Crown, White Contessa, Meigetsu, *etc.* have shown good performance in India. Australian company has released a few F_1 hybrids like Flanca, Sebal, Gigo and Nedcha. IARI, New Delhi has developed Pusa Hybrid-2 (self-incompatible line cc x Sel. 1-3-18-19). Besides self incompatible lines, genic male sterility (usually due to a recessive nuclear gene) is being used to produce hybrid varieties. Now-a-days cytoplasmic male sterility has been introduced by several sources (from *B. nigra* and *Raphanus sativus*) and is being used on commercial scale for production of F_1 hybrids. Some forms of genic male sterility are temperature unstable, resulting in possible self-pollination and contamination of F_1 seed production. Pusa Hybrid-2 is the first F_1 hybrid developed in public sector.

Development of Synthetic Varieties

Promising inbred or cultivars (6 or more) having good general combining ability can be mixed in equal amount to form synthetics (Syn_0) line. These seeds are grown and are allowed to cross together after curd formation, which will produce synthetic (Syn_1) generation. After a synthetic variety has been synthesized, it is generally multiplied in isolation for one or more generations before its distribution for cultivation. Cultivars Pusa Early Synthetic, Pusa Synthetic, Synthetic 78-1 and Pant Gobhi-3 have been synthesized from various inbred lines.

Chapter 25

Cabbage

Botanical name: *Brassica oleracea* var. *capitata*

Family: Brassicaceae

Chromosome number: 2n = 2x = 18

History of Cabbage Breeding

Cabbage is very nutritive and can be used in a variety of ways. Cabbage is a good source of minerals and vitamins A, B_1, B_2 and is an excellent source of vitamin C. Based on maturity, head shape, head size, leaf colour and shape, cabbage varieties have been classified into different groups-

- Wakefield or winningstadt group.
- Flat Dutch or drumhead group.
- Copenhagen market group (commonly grown group).
- Savoy group.
- Danish ball head group.
- Alpha group.
- Volga group.
- Red cabbage group.

In India, Copenhagen Market group and Drumhead group are commercially grown. In India, the cabbage yield is low as compared to developed countries. The reasons being short winter and lack of suitable cultivars as per climatic requirements of various agro-ecological zones and poor seed quality. Traditional cultivars perform poorly both in head quality and yield. However, the yield can be improved by developing suitable open-pollinated and hybrid cultivars resistant to diseases and for different agro-climatic regions of the country.

Botany

All these six *Brassica* species and radish (*Raphanus sativus*, 2n=18) have been intercrossed with great difficulty utilizing embryo culture. Thus the amphidiploids of fig, originated in nature from crosses between the parental species.

A cabbage flower has four sepals, four petals, six stamens in tetradynamous condition (two short and four long stamens) and a bicarpellary ovary which is superior and has a false septum. Ovules are attached on the both sides of septum. Two active nectarines are located between the bases of short stamens and ovary. The buds open under pressure of rapidly growing petals and become fully expanded in about 12 hrs. Flowers are slightly protogynous and cabbage is naturally cross-pollinated due to sporophytic self-incompatibility. Pollination is brought about by bees and flies. Bud pollination is effective to achieve selfing. For cross-pollination flower buds expected to open within 1-2 days are emasculated and are pollinated immediately with desired pollen using a brush/flower stamens.

Breeding Objectives

- ☆ To develop high yielding cultivars and hybrids.
- ☆ To develop self-incompatible but cross-compatible lines for economic hybrid seed production.
- ☆ To develop cytoplasmic male sterile lines for hybrid seed production.
- ☆ To develop cultivars suitable to grow under mild winter.
- ☆ To breed varieties resistant to black rot (*Xanthomonas campestris*) and cabbage yellows (*Fusarium oxysporum f. conglutinans*).
- ☆ To breed varieties tolerant/resistant to diamond back moth (*Plutella xylostella*), cabbage butterfly (*Pieris brassicae*) and aphids (*Breviocoryne brassicae*).
- ☆ To breed varieties for longer staying capacity in field after head formation.
- ☆ To breed varieties having short stem with narrow, short and soft core.

Germplasm Resources

Crucifer Genetics Cooperative (CrGC) is actively involved in acquiring, maintaining and distributing seed stocks and pollen of crucifer. An informative research book is published once in every three years by CrGC. A comprehensive base collection of the *Brassica oleracea* including cabbage is maintained at Horticulture Research International, Wellesbourne and is recognized by the International Plant Genetic Resource Institute (IBPGRI), Rome. Besides, germplasm information across the world is stored in a database of IPGRI. According to IPGRI policies, each sample is split into 3 parts, which are stored in the Universidad Politecnica Seed Bank, the

University of Tohokii (Sendai, Japan) and in a Seed Bank in the country where each sample was collected. Seeds can be stored in laminated aluminium foil packets, in glass jars or scaled cans at -20°C for long term storage. In India, NBPGR is the nodal agency for germplasm collection and conservation. In addition, IARI Regional Station, Katrain and Dr. Y.S. Parmar University of Horticulture and Forestry, Solan are also maintaining cabbage germplasms.

Male Sterility

Genic male sterility in *B. oleracea* is a recessive character governed by a single gene ms, a mutant from the male fertile gene Ms. Flowers and anthers of male sterile plants are slightly smaller than those of male fertiles. Two cytoplasmic male sterile cabbage lines have been developed by crossing *Raphanus* (radish) and *B. oleracea* followed by repeated backcrossing with cabbage for seven generations.

Self-incompatibility

Cabbage is self-incompatible as self-pollination produces only few seeds. The self-incompatibility in cabbage is of sporophytic type. In this system, pollen reaction is controlled by genotypes of the sporophyte or by the genome of somatic tissue of the sporophyte on which pollen grains develop. The sporophytic system of self-incompatibility can be either heteromorphic (heterostyly) or homomorphic. Homomorphic sporophytic incompatibility is controlled by one locus with multiple alleles. In cabbage, modifier genes also influence self-incompatibility (S alleles) for sensitivity to environmental factors. In the self incompatible pollination, the papilla cells of the stigma produce callose, which inhibits pollen germination and/or pollen tube growth. These SI alleles are controlled by one locus called S alleles, which are active in the stigma in flowering stage but remain inactive in the bud stage, *i.e.,* 2-4 days before its opening. The glycoprotein present in the stigma hinders the pollen germination and penetration. This phenomenon helps to maintain an individual plant or develop inbred lines for breeding work by bud pollination method.

The system of self-compatibility is characterized by following features:

- ☆ Incompatibility is controlled by one S locus having multiple alleles (50-70 alleles in *B. oleracea*).
- ☆ The reaction of pollen is determined by the genotype of the sporophyte on which pollen is produced and therefore is controlled by two S alleles.
- ☆ All the pollens of a plant have similar incompatibility reaction.
- ☆ The two S alleles may show co-dominance (independent action) or may interact by one being dominant over the other.
- ☆ The independence/dominance relationships of S alleles in pollen and in the pistil may differ.
- ☆ It is usually associated with trinucleate pollen and inhibition of pollen occurs at the stigmatic surface.

In contrast, in gametophytic system of self-incompatibility, the pollen reaction is determined by the genotype of the gametophyte, *i.e.*, pollen/egg cells itself, and in this system the pollen is binucleate and inhibition of pollen tube occurs in the style.

Various kinds of S allele interactions in a heterozygous genotype (S_1S_2) with sporophytic self-incompatibility could be as follows:

- ✰ Dominance : $S_1 > S_2$
- ✰ Co-dominance : $S_1 = S_2$
- ✰ Mutual weakening : No action by either allele.
- ✰ Intermediate gradation : 0-100 per cent activities by each allele.

Breakdown of Self-incompatibility

Young flower buds of cabbage are initially self-compatible but become self-incompatible only one or two days before the anthers mature. Bud pollination is very effective in overcoming incompatibility. Immature buds that are expected to open two days later are pollinated with mature pollen. CO_2 gas is also used to breakdown self-incompatibility. The most effective concentrations of CO_2 are between 3 and 5 per cent and the treatment duration should be between 8 and 24 hours at 100 per cent relative humidity. Spray of 3 per cent sodium chloride, half to one hour after pollination may also overcome the self-incompatibility in cole crops. Other methods of overcoming self-incompatibility are: heat treatment of stigmas with organic solvent and season pollination, steel brush pollination, double pollination and electric aid pollination but these methods are less reliable.

Self-incompatibility may be partially or wholly broken down by genetical, environmental (extrinsic) and physiological (intrinsic) factors. The genetical factors like, modifier genes, less active S-alleles and competitive interaction between S-alleles cause breakdown of self-incompatibility. Age and stage of flowering are the intrinsic factors, whereas the extrinsic factors like, high temperature, high humidity and high CO_2 concentration are responsible for breakdown of self-incompatibility.

Protogyny: Protogyny is the phenomenon where the stigma matures earlier than the anthers. In cabbage, the stigma remains receptive for five days before and four days after anthesis due to protogynous nature of the flower.

Breeding Methods and Achievements

Any breeding method applicable for cross-pollinated crops can be utilized in cabbage breeding also, however, its suitability depends on the character to be improved and mechanism to control it (oligogene or polygene).

1. **Introduction:** Earlier varieties evolved were largely introductions from abroad. They were acclimatized, tested and released for cultivation at farmer's field. Examples of such introductions include September from Germany and Copenhagen Market from Denmark.

2. **Mass selection (population improvement):** In mass selection, the plants selected from a population are allowed to flower and pollinate at random, and the seed after harvesting is pooled. This method is not much effective in breeding because there is no control over the pollen parent, however this method is still used in maintenance and purification of existing varieties. For increasing the effectiveness of selection in cabbage improvement, mass pedigree and family selection methods have been proposed. Golden Acre and Pride of India both are apparently early selections of the cultivar Copenhagen Market. Pusa Drumhead is selection from EC-6774 of Japan.
3. **Recurrent selection:** In recurrent selection, a single cultivar or a composite population is improved for yield by concentrating on additive variance. The success in recurrent selection depends upon the base population, gene frequency, intensity of selection and inter-crossing. Progeny testing and recurrent selections are based on the progeny test but in the former, pollination is at random without any control on the pollen parent whereas in the later controlled pollination is exercised.
4. **Reciprocal recurrent selection:** Reciprocal recurrent selection used two source populations to improve them simultaneously. The two populations; A and B are selected and each serves as tester for the plants selected from the other population. In this selection scheme, plants are selected for general and specific combining abilities. The resulting two populations may be intermated to produce superior populations with a broad genetic base or the inbred may be isolated from the populations to produce single or double cross hybrids or synthetics.
5. **Inbreeding:** In progeny testing (line breeding) method, the individual plant progenies are grown in rows to test the performance and breeding behaviour. The superior progeny lines showing similarity are composited by mix pollination. Inbreeding and selection has been used in cauliflower for the improvement in uniformity and vigour. Although it induces depression in vigour, but it helps to attain uniformity in head shape, size, maturity and other characters, particularly in heterozygous populations. The inbred lines developed by inbreeding can be utilized for the production of F_1 hybrids or synthetic varieties or in inter-varietal crosses for improvement.
6. **Hybridization:** Hybridization refers to the crossing of two plants or varieties of dissimilar genotype in order to improve one or more characters. The F_2 or later generations of inter varietal crosses may utilized for recombination breeding for characters involving polygenic control. Transgressive segregants (plants superior to either of the parents) in the F_2 generation may also appear. Bio-parental progeny selection in the elite F_2 generation may be helpful in producing desirable gene and improving the characters showing additive variance. Cabbage

variety Pusa Mukta is developed from an inter-varietal cross of EC 24855 and EC 10109.

Heterosis Breeding and Hybrid Seed Production

In cabbage, cultivation of hybrid varieties is becoming very popular and demand for open pollinated varieties is reducing all over the world. In 1950, Nagaoka No.1, which was the first cabbage hybrids in the world was developed through self-incompatible lines in Japan. Heterosis breeding utilizes mainly the dominance variance. The development of F_1 hybrid cultivars in cabbage is comparatively easier and cheaper method of hybrid seed production due to the presence of genetic mechanism of self-incompatibility and cytoplasmic male sterility.

Heterosis in cabbage has been documented for yield, earliness, uniformity in head size and weight and marketable head yield. Disease resistant hybrids have been produced commercially in cabbage utilizing self-incompatibility and cytoplasmic male sterility. Multiple disease resistant hybrids; Hybelle and Sanibel were also developed in Cabbage. The diallel cross has been frequently used to evaluate the magnitude of heterosis in F_1 hybrids.

Self-incompatible parental inbred lines are used for production of hybrid seeds. Inbreds are produced by bud pollination. Inbreeding helps to attain uniformity in head shape and size, maturity, especially in heterozygous strains.

The primary methods of hybrid seed production based on incompatibility are as follows:

- ☆ **Single cross hybrids:** Two self-incompatible and cross compatible inbred lines with different S alleles are planted in alternate rows or one row of parental line A for every second or third row of parental line B. The seeds are produced from both the lines. Seed produced by both the lines are identical except the characteristics, which are maternally inherited. Single cross hybrids are more uniform than double or top cross hybrids. However, the quantity of seed produced is very low, therefore it becomes expensive and hence uneconomical.
- ☆ **Double cross hybrids:** A cross between two single crosses is known as double cross. Four homozygous self-incompatible inbred lines A, B, C and D are required to produce a double cross or four- way cross hybrids [($S_1 S_1$ x $S_2 S_2$) x ($S_3 S_3$ x $S_4 S_4$)]. These lines are isolated on the basis of combining ability (gca and sca). Seeds can be harvested from every plant in the field or from both the single crosses therefore, cost of hybrid seed is reduced.
- ☆ **Top cross hybrids:** A good open-pollinated cultivar is used as pollen parent, and a single self-incompatible inbred as the female parent. The seed is harvested from female parent. In USA, most of the hybrids are top cross. The production of a F_1 cultivar using incompatibility requires 6-8 years from the initial selection of parental strains until variety release for

cultivation. In Japan, Europe and the USA the preference for F_1 hybrids of cabbage is mainly for their uniformity of harvest, non splitting trait and resistance to diseases. In India, development of hybrid varieties in cabbage still far behind as the improvement work is restricted to temperate regions only. Cabbage requires specific low temperature for chilling (mean temperature of 4-8°C for 40-60 days) to enter the reproductive phase. However, at Regional Station of IARI, Katrain developments of hybrid are in progress.

Chapter 26

Okra

Botanical name: *Abelmoschus esculentus* (L.) Moench

Family: Malvaceae

Chromosome number: 2n = 130

History of Okra Breeding

Okra or lady finger is a fast growing annual herbs and the young seed pods of which are used as a common vegetable. Recent genetic improvement has emphasized plant characteristics such as semi-dwarf plant stature, reduced branching, moderately lobed leaves for easy harvest, early maturity and smooth, dark green pod with slow fiber development. There is not much heterosis reported in okra. Most of the important yield components, *i.e.,* pods/plant, pod yield/plant, branches/plant, plant height, flowering behaviour and seed yield lack stability due to strong environmental influence, suggesting the need for breeding for specific environment. The Indian sub-continent is an important center of diversity for okra. Much effort have been made during the last five decades to enrich its variability which has also been utilized in developing promising cultivars that are adapted to a wide range of agro-climatic conditions and cropping patterns, possess better fruit yield and quality combined with disease resistance and pest resistance.

Botany

Okra is an erect herbaceous annual, become woody at maturity. Stem is green with or without radish tinge. Leaves are alternate 3-7 lobed palmate, hirsute and serrate. Flowers are solitary axillary having epicalyx (upto 10). There are 5 yellow petals with crimson spot on claw. Stamina column consist of numerous stamens which are united to the base of petals. Stigma is 5-9 lobed. Fruit is capsule. The flowering starts from below to upwards. Dehiscence usually occurs around

8.00-10.00 am, about 20 minutes after anthesis. Flowers remain open for shorter duration and wither in the afternoon. The stigma is receptive during anthesis, hence pollination is not very successful at bud stage. It is basically a self-pollinated crop, however, cross-pollination through insects can be as high as 19 per cent hence okra is classified as often cross-pollinated vegetable crops.

For emasculation, buds likely to open the next day are given a slight ring cut at the base of the flower bud with the help of a blade. Petals along with calyx sheath are removed and stamina column and stigma are exposed. The indehiscence anthers are removed with a pair of forceps. The emasculated buds are bagged. Next morning the bags are removed and the emasculated buds are pollinated with the male flowers bagged a day earlier. Pollinated flowers are bagged and labeled. After a few days the bags may be removed.

Germplasm Resources

Germplasm base in okra is relatively narrow. The maximum variation is observed in the widely adopted cultivated species, *Abelmoschus esculentus* which is also called the Saudanian type. Turkey and India are maintaining maximum collections of this group. This group is also distributed in Afghanistan, Iran, Pakistan, Bangladesh and Yugoslavia to a lesser extent. The other group, Guinean is found in West Africa and is distinguished from other groups on the basis of red leaf veins, very late flowering, reduced number of sub-calyx bracts and more number of seeds per pod. It is considered to be natural amphidiploids of *A. esculentus* and *A. manihot*. In India, the major okra collection is maintained by the NBPGR, New Delhi, and IBPGR has assigned it global responsibility for base collection. Okra germplasms endemic to India and Bangladesh have been given priority for collection and evaluation. Large numbers of genotypes, including wild relatives have been collected and are being made available for further testing and utilization. So far, the primary objective has been to identify high yielding genotypes carrying resistance to yellow vein mosaic and okra leaf curl virus. Besides, long fruiting habit, resistance to various other biotic stresses, *viz.*, shoot and fruit borer, jassid and high vitamin C content present in the wild species can be utilized for the improvement of *A. esculentus*. Germplasm collection from Asia and Africa should be augmented for broadening the genetic base of okra for direct breeding found in West Africa and is distinguished from other groups on the basis of purposes as well as for conservation of the germplasm.

Breeding Objectives

- ✰ To evolve high yielding varieties with wide adaptability.
- ✰ To develop cultivars of different maturity group suited to specific environment.
- ✰ To develop multiple disease and pest resistant varieties with special emphasis on yellow vein mosaic virus, okra leaf curl virus, shoot and fruit borer, jassid, leaf hopper and root knot nematode.

- ✰ To evolve varieties tolerant to abiotic stresses, especially to low temperature, drought, excessive rain, saline and alkaline soils.
- ✰ To develop varieties having dark green, tender and smooth pods suitable for export markets.
- ✰ To evolve varieties with better nutritive quality and also suitable for dehydration, canning and freezing.
- ✰ To develop high yielding and disease resistant hybrids.

Breeding Methods and Achievements

Okra is often cross-pollinated crop. Breeding methods adapted to self-pollinated crops can be employed for varietal development in okra. The methods commonly employed are; plant introduction, pureline selection and hybridization using back-cross technique, mutation and polyploidy breeding.

1. Introduction

The cultivar Pusa Sawani bred as a symptomless carrier of yellow vein mosaic virus has been introduced to the different okra growing regions of the world and is widely cultivated. Similarly, the varieties Clemson Spineless, Louisiana and Velvet Round bred in the USA have been introduced to a large number of okra growing countries both for fresh market and canning purposes. Indigenous introductions have played a significant role in improving the okra wealth of many countries. *Abelmoschus manihot* sub species manihot now known as *A. coillei*, a semi wild species introduced from Ghana has served as a source of resistance to yellow vein mosaic virus.

2. Pureline Selection

In India, the first improved variety of okra, *i.e.,* Pusa Makhmali' was developed through pureline selection from material collected from West Bengal. The cultivars CO-1 is also a single plant selection from a heterozygous population of 'Red Wonder'. Gujarat Bhindi 1 is a pureline selection from an unknown bulk seed.

3. Hybridization

In India, inter-varietal hybridization followed by pedigree selection produced the widely cultivated, high yielding, yellow vein mosaic virus (YVMV) tolerant cultivar 'Pusa Sawani'. However, this variety has become susceptible to YVMV in plains, but it is doing well where YVMV does not appear. Similarly, Selection 2 is a derivative of multiple inter-varietal crossing. Inter-specific hybridization has been followed in the development of Punjab Padmini, Winter Bush, Prabhani Kranti, P-7, Arka Anamika and Arka Abhay. In all these cultivars, except Punjab Padmini, backcrosses were made.

4. Mutation Breeding

Efforts have been made to generate more variability through mutations. A single dominant gene controlled the mutant. A mutant EMS 8 carrying resistance to YVMV and tolerance to fruit borer has been evolved through the use of ethyl methanesulfonate. In large M_1 and M_2 population, male sterility has been observed. Besides, variations in number of ridges, *i.e.,* 4-11 are also observed.

5. Polyploidy Breeding

Failure of inter-specific crosses due to hybrid sterility prompted researchers to try amphidiploids. A fertile amphidiploid has been reported by duplication of the F_1 between *A. esculentus* x *A. manihot* called Nori-Asa. An amphidiploid, *A. tuberculata* x *A. esculentus* has also synthesized. An amphidiploid between *A. esculentus* and *A. manihot* subsp. *manihot* has also been reported. This amphidiploid was resistant to yellow vein mosaic virus. In 1983, successful attempt was also made in inducing polyploidy in three inter-specific hybrids. Amphidiploidy in a cross between *A. esculentus* and *A. manihot* subsp. *tetraphyllus* has also been reported.

6. Heterosis Breeding

Over-dominance observed for days to first flowering, fruit number, length and weight, plant height and yield, and can be exploited commercially for manifestation of heterosis. The potential of this approach is evident from reports of more than 100 per cent increase in yield over the better parent and also the presence of dominance gene effects for yield and yield contributing characters. Moreover, hybrid vigour can also be utilized for developing hybrids carrying resistance to yellow vein mosaic virus, as the resistance is dominant over susceptibility. For production of hybrids, inbreeding for two or three generations are sufficient to develop inbred lines. These inbred are used in hybridization by hand emasculation and pollination. The phenomenon of male sterility and protogyny though reported is not commercially exploited so far. In India, a number of hybrids have been developed both by private and public sector organizations.

Breeding for Quality Traits

The important characteristics that determine the quality and processing ability are; yield, fruit number, length, weight and colour, pod pubescence and spines, fiber, vitamin C and protein, total sugar content, and resistance or tolerance to pests and diseases. Improvement of various quality and processing attributes of okra have never been attempted through direct breeding. The improvement, if any in these traits has come through breeding for other economic characters. Now, it is imperative to start the work on quality breeding to enhance export possibilities.

Chapter 27

Carrot

Botanical name: *Daucus carota* L. var. *sativa* Hoffm.

Family: Apiaceae (Umbellifereae)

Chromosome number: 2n = 2x = 18

History of Carrot Breeding

Nutritionally, carrot is a very important crop for mankind. However, systematic carrot breeding work is lacking in our country. IARI, Regional Station, Katrain started carrot breeding in 60s and few varieties, *viz.*, Pusa Kesar, Pusa Meghali and Pusa Yamdagni were evolved. Mass selection and pedigree selection within different populations have been used to develop carrot varieties. Generally, open pollinated varieties are adapted to different situations. Nevertheless, all efforts to breed varieties with high level of uniformity have been limited by genetic factors originating from breeding methods used for open pollinated varieties. A major reason for this is the heterozygosity of the open pollinated varieties, while other is the inbreeding depression resulting from random selfing of plants within such populations. All open pollinated varieties suffer from some degree of inbreeding depression and have a limited degree of uniformity. In cases, where extreme uniformity seems to be unnecessary, *e.g.*, for juice or pulp production or for regions where quality is not so desired, breeding of synthetic varieties would also be worthwhile.

Botany

The inflorescence of carrot is a compound umbel. A primary umbel can have over 1000 flowers at maturity, whereas secondary, tertiary and quaternary umbels bear fewer flowers. Floral development is centripetal, *i.e.,* the flowers to dehisce first are on the outer edges of the outer umbellets. Carrot is protandrous. After straightening of filament, the pollen is shed and stamens quickly fall. After this, the petals open fully and the style elongates. The style is divided into two parts. The petals of petaloid plants are persistent unlike those of brown-anther, male sterile plants. Flowers are epigenous. There are five small sepals, five petals, five stamens and two carpels.

Emasculation is laborious and time consuming. As soon as the first bud in umbel opens, the whole umbel of the female parent is bagged in a muslin/cloth bag. The flowers are removed daily until peak flowering has reached. Anthers are removed from the early opening outer flowers in the outer whorl of umbellets until sufficient flowers are emasculated. Unopened central florets in the emasculated umbellets and all late flowering umbellets are removed. Thus, only the emasculated flowers are left on the female inflorescence inside the bag. A pollen bearing umbel from previously protected male plant is inserted into the bag of the female parent along with some house-flies to ensure pollination. Daily for a few days in the morning, the male umbel is gently rubbed against the emasculated umbel to enhance artificial cross-pollination.

Sometimes, 1-2 flowering umbels of both the parents are enclosed in the same cloth cage along with some houseflies. Seed from each parent is sown in adjacent rows. The hybrids and the parents could be identified (not always) and necessary roguing done to remove the selfed plants.

Breeding Objectives

- Early root maturity.
- Reasonable root yield with high root: shoot ratio.
- High anthocyanin content (red coloured roots) in Asiatic types and high β-carotene content (deep orange coloured roots) in European type carrot.
- Uniformity in root shape and size.
- Thick flesh having thin and self coloured core in roots.
- Cylindrical, uniformly tapering or stump rooted carrot with non-branching habit.
- Smooth root surface.
- Cracking and forking free roots.
- High sugar and dry matter contents in roots.
- Slow bolting habit.
- Resistance to biotic and abiotic stresses.

Germplasm Resources

The major carrot germplasm collections are maintained at USDA, Plant Introduction Station Ames, Iowa and US National Plant Germplasm System, Griffin, USA and Chinese Crop Germplasm Information System (CGRIS), China. In India,

NBPGR is the nodal agency for germplasm collection and conservation. Temperate type carrot germplasms are maintained at IARI, Regional Station, Katrain and Dr. Y.S. Parmar University of Forestry and Horticulture, Solan, H.P.

Selection Criteria in Carrot

Root length: Total yield increases with increase in root length, but it should not usually exceed to 25-30cm, because the roots longer than this will be more liable to damage during harvest and handling.

Colour and quality: Visual examination of roots, cross-section of roots and longitudinal section of roots are effective tools in deciding the colour and quality of carrot. There should be uniform colour in whole root. The colour of xylem, phloem and vascular cambium should match as far as possible.

Sugar and flavour: Total sugars which contribute to sweetness are an important component of general preference. Sugars can be estimated by a refractometer. Selection for high soluble solids and dry matter is also possible by measuring specific gravity. High dry matter is useful for processing industry. Sugar and flavour of carrot roots can be judged by organoleptic test also. Phenotypic recurrent selection has been found successful to improve the quality characteristics of carrot.

Non-bolting: Bolting may cause serious losses in yield and quality, hence it is important to apply selection pressure for non-bolting.

Disease resistance: Susceptible cultivars are generally planted between rows of breeding lines, and the spreader row plants are inoculated to ensure the spread of disease. This practice is applicable for both Alternaria leaf blight and *Cercospora* leaf spot.

Emasculation and pollination: Emasculation is laborious and time consuming job. As soon as the first bud in an umbel of female parent opens, the whole umbel is bagged with muslin cloth. Anthers are removed from the early opening outer flowers in the outer whorl of umbellets until sufficient flowers are emasculated. Unopened central florets in the emasculated umbellets and all late flowering umbellets are removed. Thus, only the emasculated flowers are left on the female inflorescence inside the bag. A pollen bearing umbel from previously protected male plant is inserted into the bag of the female parent along with some houseflies to ensure pollination. The male umbel is gently rubbed against the emasculated female umbel daily for a few days in the morning to enhance artificial cross-pollination. Sometimes, 1-2 flowering umbels of both the parents are enclosed in the same cloth cage along with some houseflies. Seed from each parent is sown in adjacent rows. The hybrids and the parents could be identified and necessary roguing may be done to remove the selfed plants.

Breeding Methods and Achievements

Mass Selection

It is practiced to purify an old or genetically mixed variety. The superior and uniform plants are selected and uprooted. Plants are observed for root and foliage characters. Besides, phenotypic characters of roots, the cortex and core texture and their colour are also visually observed for selection. Roots are also tasted for edible quality. Selected roots are planted in field in isolation and their seeds are bulked together for growing in next season. The process is repeated till desired improvement is achieved. The genotype is tested against standard check, and if found superior, is recommended for commercial production.

Hybridization

Inbreds are developed by simple phenotypic recurrent selection. Crosses are made between two parents and in later generations combined mass pedigree selection is practiced. Varieties such as, Pusa Kesar, Pusa Meghali, Sel-233 and Pusa Yamdagini have been developed by hybridization.

Heterosis Breeding

Poole (1937) was perhaps the first person who reported heterosis in carrot, while Thompson (1961) and Hanschke and Gabelman (1963) were the first to detect and analyze male sterility. The first carrot hybrids were available in early 1960s in the USA. Today, more than 100 hybrid varieties exist worldwide. The percentage of hybrids is 60-90 per cent in Europe for early and late varieties and in USA, has reached almost 100 per cent.

Sterile cytoplasms have been identified, which condition either petaloid (*pt*) or brown anther (*ba*) male sterility with different genetic backgrounds and origin. The brown anther type (*ba*) is present in all cultivated orange coloured open-pollinated varieties. The phenotype is characterized by deformed, brown coloured anthers without functional pollen caused by a genetic block in meiosis. The brown anther type (*ba*) results from an interaction of the *Sa* cytoplasm with two independent nuclear genes (homozygous *aa* or dominant *B*). The two complementary genes *E* and *D* operate with their dominant alleles as restorer genes. Due to this complex inheritance, many test crosses are necessary for the development of a suitable maintainer.

The cytoplasm of the petaloid type (*pt*) is derived from a wild form of *Daucus carota* L., and has been introduced into many open-pollinated varieties of cultivated carrot. The '*pt*' type is characterized by transformation of anthers into petals or petal like structures which are unable to produce functional pollen. The petaloid type (*pt*) results from an interaction between Sp-cytoplasm and two independent dominant genes (M1 and M2). A maintainer for this type cannot be detected in the F_1, because of the dominant state of M genes. Backcrosses should be performed by use of the male fertile genotype (*Sp*) m1m1, m2m2 as a tester in order to find dominant alleles.

A third CMS system has been detected in an alloplasmic form of orange coloured carrot originating from a cross between the wild carrot *D. carota* gummifer Hook and the cultivated carrot *D. carota* L. sativus Hoffm. This type of male sterility called '*gum*' type is characterized by a total reduction of anthers and petals. Recent results on the genetic mechanisms suggest that an interaction of the gummifer cytoplasm with a recessive allele (*gugu*) in the nucleus is responsible for the expression of this type of male sterility.

The two CMS system '*ba*' and '*pt*' generally suffer from instability of male sterility influenced by high temperature. Nevertheless, observations over many years have revealed that other factors such as dry conditions, growing time or long-day conditions operate provocatively. A strict selection scheme is therefore, necessary because carrot is partially andromonoecial, *i.e.*, in umbels of higher order male flowers can be produced. Umbels of the 5-7th order must be examined carefully.

The development of the ms and maintainer lines is very laborious due to the dominant state of male sterility. Crossing, back crossing, selfing and testing of the progenies in the following two generations, including isolation of the positive progenies are salient steps in breeding process. The breeder is forced to eliminate male fertile plants or phenotypes with partial male fertility within the ms lines developed, and to develop new lines with low inbreeding depression.

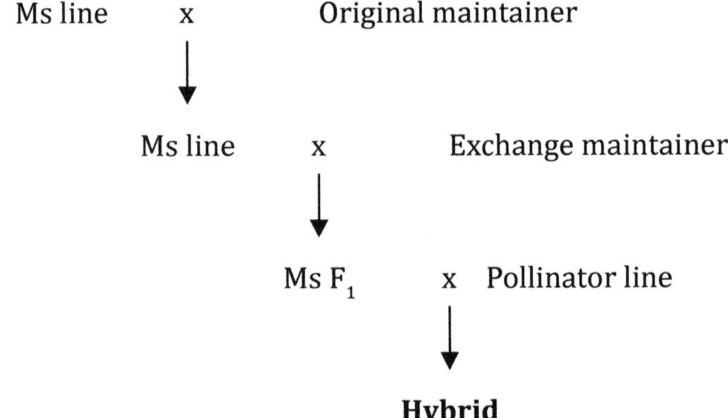

Chapter 28

Onion

Botanical name: *Allium cepa*

Family: Amaryllidaceae

Chromosome number: 2n = 2x = 16

History of Onion Breeding

Onion is an important vegetable crop grown throughout the world. Onion is indigenous to India. Natural variability is created constantly due to its outbreeding nature. Therefore, adoption and acclimatization of onion has undergone continuous selection by onion growers to suit as per climate and market demands. Local type cultivars like Nasik Red, Poona Red, Patna Red and Bellary Red are the result of selection by growers. Systematic breeding programme was started as early as 1960 at Pimpalgaon Baswant, Nasik and later on at IARI, New Delhi.

Two types of onion are commercially grown in India:

1. **Common onion** (*Allium cepa* var. *cepa*)- This is most important in commercial trade. Bulbs are large, normally single and plants reproduce through seeds.
2. **Multiplier onion** (*Allium cepa* var. *aggregatum*)- This produces bulbs of smaller size and they are several in number to form an aggregated cluster. Propagation is usually vegetative via daughter bulbs.

The programme was further strengthened under coordinated project through SAUs and ICAR institutes. As a result, a number of varieties of common onion and multiplier onion have been developed. Introduction, mass selection, selfing and massing, inbreeding, hybridization and heterosis breeding have been used for improvement of onion.

Botany

The flowering stalk of onion is an apical extension of the stem but without nodes and internodes. The growth of flower stalk ceases when umbels start to flower. The length of scape (flower stalk/seed stem) is controlled by genetic factors, long stalks Dwl, being dominant over dwarf, dwl stalks. Spherical umbel which terminates the stalk may have 50-2000 flowers, common range being 200-600. Inbreds produce significantly less flowers per head than heterozygous plants. Each flower is attached to a slender pedicel. Secondary umbel (those arising from branches that have already flowered) have approximately 30-50 per cent of the number of flowers in the primary heads.

The flower stalk of onion reaches to a final length of 1-2m. The growth of flowering stalk is stopped when umbel start flowering. The flowering structure is called an umbel which is an aggregate of many small inflorescences (cymes) of 5-10 flowers, each of which opens in a definite order causing flowering to be irregular and to last for 2 or more weeks. Each individual flower contains 6 stamens, 3 carpels united into one pistil and 6 perianth segments. The pistil contains 3 locules each of which has 2 ovules.

Breeding Objectives

- High yield.
- Superior bulb quality traits (size, shape, colour, pungency, firmness and dormancy), high total soluble solids content (important for dehydration industry), skin retention and high dry matter.
- Resistance to diseases (purple blotch, basal rot, stemphyllium blight, anthracnose, pink root rot and bacterial rot).
- Resistance to insect pests, mainly thrips.
- Resistance to abiotic stresses (moisture stress, high temperature, salinity and alkalinity).
- Development of high yielding varieties capable of producing good seed yield.
- Development of disease resistant F_1 hybrids with superior quality bulbs.
- Development of varieties suitable for export market.

Characteristics of an Ideal Bulb

Bulb should be attractive, uniform in size, shape, colour and time of maturity. It should consist flavour (pungency) as per requirement, high dry matter and long storage life. Additional desirable features are intact and attractive skins, thick leaf scales (rings), single centered bulb, thin neck and resistant to early bolting, diseases and pests. Each of these characteristics is genetically inherited but can be modified by environmental and cultural practices.

Germplasm Resources

The following centres in India are maintaining the onion germplasms/lines:

- ☆ Directorate of Onion and Garlic Research (DOGR), Rajgurunagar, Pune.
- ☆ National Bureau of Plant Genetic Resources (NBPGR), New Delhi.
- ☆ Indian Institute of Horticultural Research (IIHR), Bangalore.
- ☆ Mahatma Phule Krishi Vidyapeeth, Rahuri.
- ☆ National Horticultural Research and Development Foundation (NHRDF), Nashik.
- ☆ Indian Agricultural Research Institute (IARI), New Delhi.
- ☆ Punjab Agricultural University, Ludhiana.
- ☆ Tamil Nadu Agricultural University, Coimbatore.
- ☆ Central Institute of Temperate Horticulture, Srinagar.
- ☆ Vivekananda Parvatiya Krishi Anusandhan Shala, Almora.

Besides, the germplasm lines are also maintained in USA, European Community, Japan and Israel.

Selection Techniques/Criteria

- ☆ **Yield:** Evaluation at uniforn density.
- ☆ **Maturity:** Time of foliage collapse.
- ☆ **Bolting:** Good assessment by early sowing.
- ☆ **Quality assessment:** Visual, colour, skin retention, neck thickness, shape and uniformity.

Widely recognized phenotypic correlations between large size, softness, low pungency and poor storage ability should also be taken into account.

- ☆ **Storage capacity:** Dormancy is important because onions are normally stored for longer time. By storing bulbs in environmental conditions close to those under which a commercial crop would normally be stored and subsequently recording the bulbs for-marketability, losses due to rotting and sprouting.
- ☆ **TSS:** High TSS is important for dehydration industry producing onion chips and powder.

Quantitative Traits and Bulb Colour

The major quality attributes are size, shape, colour, pungency, firmness and soluble solids (carbohydrate content). Bulb shape may be round, pyriform or flat, soft or firm, and white, red brown or yellow colour. Long dormancy is desired for storage. Likewise, higher soluble solid are valuable for dehydration in the form of onion chips and powder. The degree of pungency is variable factor for customers and depends on the concentration of volatile sulphur compound. The pungent onions have more storability than the non-pungent one. The onion having low moisture at maturity, resulting in firm flesh, higher concentration of solids and pungency compound per unit dry weight and hence longer storage life. Most of the quality characteristics is governed genetically and can he manipulated by suitable breeding strategies. However, some of the characters depend upon the environment. Bulbs subjected to stress during the growing season are more likely to sprout in storage.

The characters like bulb yield, bulb shape, number of leaves, TSS, dry matter content, umbel stalk height and seed weight are controlled by both additive and non-additive gene action (polygenic), except few contradictions. Four colour classes, *i.e.*, white, yellow, red and brown are controlled by a series of complementary epistasis genes with a basic colour factor (C) necessary for the development of red or yellow.

Breeding Methods and Achievements

Since onion is a cross-pollinated crop, all the methods employed for the improvement of the cross-pollinated vegetables can be employed for the improvement of the onion. Products of these improvement strategies are open-pollinated varieties or hybrids.

Breeding of Open-Pollinated Varieties

Open-pollinated varieties are defined as genetically variable populations which are maintained and multiplied by mass pollination in isolation.

1. Introduction

Early Grano and Brown Spanish were introduced and acclimatized for growing in Indian conditions. Brown Spanish is a long-day type variety grown on hills.

2. Mass Selection (Population improvement)

The simplest population improvement method applicable to onion is mass selection in which 1-5 per cent of mother bulbs of the desired characteristics are selected from a chosen basic source population at harvest or after storage. Later on, these selected bulbs are replanted in isolation to produce mass pollinated seeds. These steps could be repeated till substantially improved new population (variety) is produced. The success of this method depends upon:

- Genetic variability in the initial population, and
- High heritability of the character under selection.

A modification of this mass selection is stratified mass selection, in which the field plot of the bulb crop is subdivided into equal sized area and mass selection is carried out into each area to make allowance for the effect of variable growing conditions within the field.

More refined open-pollinated breeding schemes are based on family selection under the situations where heritability's are usually low. In these schemes, selection is based on family mean performance rather than the individual bulb as the unit of selection. The families could be half-sib families or S_1/S_2 families. Population can be improved by recurrent selections on the basis of phenotypic performance. Even in seed production, selection of bulb (mass selection) is necessary to avoid deterioration in cultivar.

3. Mutation Breeding

A field of mutation breeding has greatly changed with an introduction of artificial transmition of gene by X-rays. Radiations and chemical mutagens are the potential mutagenic agents for induction of mutations. In onion chemical mutagens *viz.*, ethyl-methane sulphonate, N-methyl-N-nitrosourea and ethyl-imine along with gamma rays have been used.

4. Breeding of Hybrids

The development of an F_1 in onion requires the development of a male sterile line (A line), an inbred maintainer for A line (B line) and a pollinator male parent (C line). These lines have the following genetic constitution:

A line = S msms

B line = N msms

C line = N MsMs

Chapter 29

Garden Pea

Botanical name: *Pisum sativum* L.

Family: Fabaceae (Leguminoceae)

Chromosome number: 2n = 2x = 14

History of Garden Pea Breeding

The pea is cultivated since the ancient period. In garden pea, early varieties are more remunerative due to higher market price in the beginning of the season. Pod colour, size, attractiveness and seed size are also considered as premium traits. Therefore, breeding programme should be developed for short duration varieties which can fit in the multiple cropping/relay farming system and can be grown under a wide range of agronomic and environmental conditions.

An ideotype for garden pea should have higher number of pod/plant number of seed/pod and high shelling percentage. Other attributes associated with green pod yield and qualities are: early flowering, short height (60-90cm) short internodes while retaining the same number of internodes and number of branches. However the main features of an ideotype in drought conditions are reduced canopy with high photosynthetic rate. In peas substantial variability is reported for nodulation that influenced nitrogen fixation. Therefore, specific conditions should be selected for maximum symbiotic fixation.

Botany

Pea is an annual herbaceous plant. It has a tap root system. Stems are slender, usually single and upright in growth. Leaves are pinnately compound with two to several leaflets. The rachis terminates in a simple or branched tendril. There are large stipules at the base of leaf. The plant may be single stemmed or many axillary stems may originate at the cotyledonary node or any superior nodes, specialy if the apical growing point is destroyed, leaflets of a pair are opposite or slightly alternate. The lower leaflets are larger than the upper leaflets. The margins of leaflets and stipules may be entire or serrated.

The inflorescence is raceme arising from the axil of a leaf. The lowest node at which flower initiation occurs in normally constant under a given set of conditions and is used in classifying the varieties into early and late type. Most early cultivars produce the first flower from nodes 5-11 and the late cultivars start flowering at about nodes 13-15. The flowers are typically papilionaceous with green calyx comprising of five united sepals, five petals (one standard, two wings and two keels). The stamens are in diadelphous (9+1) condition. Nine filaments are fused to form a staminal tube while the tenth is free throughout its length.

Pea is strictly self-pollinated in nature. Stigma is receptive to pollen from several days prior to anthesis until one day or more after the flower wilts. Pollen is viable from the time anthers dehisce until several days thereafter. For emasculation the flower bud chosen should have developed to the stage just before anther dehiscence, indicated by extension of petals beyond sepals. Flowers can be emasculated at any time.

Germplasm Resources

Germplasm is the basic requirement to develop new varieties. It consists of wild species, landraces, old varieties, new varieties and advanced lines. The following are the germplasm resource centres at global level:

- ✰ NBPGR - National Bureau of Plant Genetic Resources, New Delhi, India.
- ✰ IIVR- Indian Institute of Vegetable Research, Varanasi, India.
- ✰ IIHR- Indian Institute of Horticulture Research, Bangalore, India.
- ✰ ICRISAT- International Crop Research Institute for Semi Arid Tropics, Hyderabad, India.
- ✰ IIPR- Indian Institute of Pulse Research, Kanpur, India.
- ✰ Nordic Gene Bank, Landskrone, Sweden.
- ✰ North East Regional Plant Introduction Station, Geneva, New York.
- ✰ The Weibullsholm Pisum Genetica Association collection for Kulturplflanzenforschung, Gattersleben, Germany.
- ✰ The Labaratoro Del Germplasm, Bari, Italy.

Breeding Objectives

- ✰ High green pod yield.
- ✰ Long to medium sized, well filled, attractive green pods with more seed/ pod.
- ✰ High sweetness.

- ☆ High shelling percentage.
- ☆ Specific maturity group (early, mid, and late).
- ☆ Suitable freezing and canning.
- ☆ Resistance to biotic and abiotic stresses.
- ☆ Breeding cultivars for high nutritive value (rich in protein, vitamins, sugars, and minerals).
- ☆ Breeding for processing traits (canning, freezing, and dehydration).

Breeding Methods and Achievements

Introduction and Selection

Garden pea is a new introduction in our country. A number of cultivars have been introduced from various European countries and USA through NBPGR, New Delhi. The introduced cultivars were tested for their acclimatization to local environment and those found suitable were released for cultivation. In some cases, plants selected from variable populations were released as cultivars. A substantial number of superior introductions are being used in hybridization programme to transfer desirable characters. Some of the well known introductions are Arkel, Early Badger, Meteor, Bonneville, Multifreezer, Little Marvel and Perfection New Line.

Hybridization

For crossing, emasculation is done by exposing the keel with forceps in the afternoon and pollination in the following morning taking viable pollen with the help of forceps or needle. The pollination may be performed on the same day after emasculation. A good pod setting of 80- 100 per cent has been observed under ideal condition.

Generally, inter-varietal hybridization is practiced as a breeding method to combine good agronomic traits of two varieties. Mostly single crosses have been attempted. The emphasis has been given to develop dwarf or medium dwarf and early maturity types. After hybridization between two desirable parents, pedigree and bulk, single seed descent (SSD) selection procedures are followed. SSD method is now becoming common practice in peas. This is particularly useful in the situation where selected better lines are intercrossed. F_1 plants are grown to produce 500 or more F_2 seeds. One seed is harvested from each F_2 plant and harvested seeds are bulked to plant F_3. This procedure continues till F_5, in which phenotypically superior individual plants are selected for progeny planting and evaluation. This method requires less resources and rapid advancement of generation is possible in field and glasshouse/off-season nursery.

Mutation Breeding

Physical (gamma rays) and chemical (EMS, NEU, EI, NMU and Sodium azide) mutagens can be used for varietal development in pea. A dose of 10-15

Krad of gamma rays is used for irradiation of seeds. A good judging criterion of effectiveness for any mutagen is germination reduction, not exceeding 50 per cent germination for gamma radiation, but only 30 per cent for neutrons and certain chemical mutagens. High germination reduction may result due to more number of chromosomal aberrations, and this in turn will lead to high sterility. Chemical mutagens are applied as water solutions and seeds are usually pre-soaked for 12-16 hours. Pre-soaking facilitates the penetration of the mutagen into the tissues. The optimum temperature and duration are 21-24°C and 2-4hr. respectively. Recommended concentrations of chemical mutagens are; EMS (0.05-0.30 per cent), NEU (0.20-0.40 mmol), EI (0.05-0.15 per cent), NMH (0.01-0.03 per cent) and DES (0.03 per cent). The solutions should be prepared fresh and buffered to pH 5-6.

Approximately 1000 fertile plants are the minimum requirement for the M_1, generation. Effective mutagen treatments may lead to only one mutation per locus in 100000 cells. M_1 generation should be grown under optimum condition. Each M_1 plant should be harvested individually for growing M_2 progenies. If M_1 generation is large, a single pod or even a single seed/M_1 plant may be harvested. Starting with the M_2 generation the breeding methods are very similar to those used for hybridization programme from F_2 onwards. Most reliable method is pure pedigree selection. Phenotypically most mutations are not expressed unless homozygosity is reached, *i.e.,* not earlier than the M_2 generation. Segregation in M_3 and later generations gives reliable information on the inheritance of induced mutations.

Chapter 30

Cucurbits

History of Cucurbits Breeding

The family cucurbitaceae is named from genus *Cucurbita*. The vegetable crops belonging to the family cucurbitaceae are known as cucurbits. The plants are generally herbaceous, climbers or trailers with tendrils and flowers are commonly unisexual, pistillate with inferior ovary. Cucurbitaceous crops are generally grown through seeds but some semi-perennial crops are perpetuated by vegetative means. The crop improvement methods of seed sown and vegetative propagated crops have been discussed separately.

Cucurbitacin or triterpenoid are found in cucurbits. There are at least 19 known cucurbitacin are isolated and majority of the species contain bitter principle in some portion of the plant. When bitter pollen fertilizes non-bitter ovules resulting fruit will be bitter. This phenomenon is known as metaxinia following species are under cultivation:

Hindi Name	Common Name	Botanical Name	Chromosome No. (2n)
Khira	Cucumber	*Cucumis sativus* L.	14
Kharbuza	Musk melon	*Cucumis melo* L.	24
Kakri	Longmelon	*Cucumis melo* L. var. utilissimus	24
Tarbuz	Watermelon	*Citrullus lanatus* L.	22
Tinda	Round melon	*Citrullus lanatus* L. var. fistulosus	24
Phoot	Snapmelon	*Cucumis melo* L. var. momordica	24
Sitaphal	Pumpkin	*Cucurbita moschata* L.	40
Chhappan Kaddu	Summer squash	*Cucurbita pepo* L.	40
Squash	Winter squash	*Cucurbita maxima* L.	40

Hindi Name	Common Name	Botanical Name	Chromosome No. (2n)
Lauki/Ghiya	Bottle gourd	*Lagenaria siceraria* L.	22
Karela	Bitter gourd	*Momordica charantia* L.	22
Kali Tori	Ridge gourd	*Luffa acutangula* L.	26
Ghia Tori	Sponge gourd	*Luffa cylindrical* L.	26
Parwal	Pointed gourd	*Trichosanthes dioica* L.	22
Chichinda	Snake gourd	*Trichosanthes anguina* L.	22
Petha	Ash gourd	*Benincasa hispida* L.	24
Kundru	Ivy gourd	*Coccinia grandis* L.	24
Chayote	Chow-chow	*Sechium edule* L.	24

Improvement of Seed Sown Cucurbits

Major crops under this group are ash gourd, bottle gourd, bitter gourd, cucumber, musk melon, pumpkin, watermelon, winter squash and summer squash, *etc.* The following methods are employed for improvement in these crops.

Breeding Methods and Achievements

Introduction

When some of the elite lines (or variability) or any species are imported from outside the country for direct as well as indirect use is known as plant introduction. A well known commercial variety or breeding line introduced from outside can be utilized directly after multi-locational trials as a variety or it can be used after proper selection. Some of the introductions of cucurbits which have been released in India after acclimatization and evaluation.

Varieties Developed through Introduction

- ☆ **Watermelon**: Asahi Yamato (Japan), Sugar Baby (USA), New Hampshire (USA), Midget Improved Shipper (USA), Dixilee (USA), *etc.*
- ☆ **Cucumber**: Japanese Long Green (Japan), Straight Eight (USA), Poinsette (USA), *etc.*
- ☆ **Sumner squash**: Australian Green (Australia), Patty Pan (USA), *etc.*
- ☆ **Bitter gourd**: MD-4.

Selection

Cucurbits being highly cross-pollinated crops have high degree of heterozygosity. Therefore, selfing followed by selection improves the population and helps to develop new variety with distinct features from original. Different selection methods can be applied in cucurbits for population improvement.

Mass Selection

Superior plants are selected from the base population and their seeds are mixed for raising the next generation. This selection procedure is repeated in uniform growing conditions up to the selection as a new variety/type. Mass selection is effective in improving the sugar content of muskmelon and watermelon. This method is effective in improving simple inherited and highly heritable qualitative characters.

Single Plant Selection

Single plant selection is very common method of selection in cucurbits. The selfed fruits of selected plants are taken from the heterozygous and non-uniform material. The seeds of selected plants are raised individually in next generation for evaluation and maintained by selfing. The cucurbitaceous crops do not show much loss of vigour due to inbreeding. Therefore, homozygosity for the concerned characters can be attained in the progeny of the individuals by selfing, after necessary evaluation, the best selection is treated as new type. Several varieties have been developed by single plant selection in cucurbitaceous vegetable crops.

Varieties Developed through Selection

- **Ash gourd:** Bindu, Kashi Dhawal, CO-1, *etc.*
- **Bottle gourd**: Pusa Summer Prolific Long, Pusa Summer Prolific Round, Punab Long, Arka Bahar, Pusa Naveen, Punjab Round, Narendra Rashmi, Kashi Bahar, *etc.*
- **Bitter gourd:** Coimbtore Long, Pusa Do Mausami, Arka Harit, Priya, Pusa Vishes, Kalyanpur Sona, Kalyanpur Barahmasi, NDB-1, CO-1, *etc.*
- **Cucumber:** Sheetal, Swarna Ageti, Swarna, Poorna, Swarna Sheetal, *etc.*
- **Muskmelon**: Kashi Madhu, Durgapur Madhu, Arka Rajhans, Arka Jeet, Pusa Madhuras, *etc.*
- **Watermelon**: Durgapura Meetha, Durgapura Kesar, *etc.*
- **Pumpkin**: Kashi Harit, Ambli, Narendra Amrit, Narendra Agrim, CO-1, CO-2, *etc.*
- **Longmelon**: Arka Harit, Karnal Selection, *etc.*
- **Ridge gourd**: Pusa Nasdar, CO-1, CO-2, *etc.*
- **Sponge gourd:** Pusa Chikni.

Inbreeding and Selection

Inbreeding is the mating of closely related individuals, *i.e.,* either selfing or sib mating. Individual selection of inbreds are practiced after attaining maximum uniformity and claimed as a variety. Few varieties of muskmelon and pumpkin have been developed by this method.

Varieties Developed through Inbreeding and Selection

- ☆ **Muskmelon:** Hara Madhu, Kashi Madhu, *etc.*
- ☆ **Pumpkin**: Pusa Viswas, Arka Chandan, *etc.*
- ☆ **Ash gourd**: Kashi Dhawal, IVAG-90, *etc.*

Hybridization

Hybridization creates new genetic variability in the F_2 and subsequent generations which helps efficient selection of desirable types. The hybrid progenies are advanced in the subsequent generations by selecting desired plant type and their selfing. Single plant selection by maintaining the pedigree of segregating generation is applied. Several varieties have been developed by hybridization and selection from advanced generations.

Varieties Developed through Hybridization

- ☆ **Bottle gourd:** Punjab Komal (LC-11 x LC-5), Pusa Naveen, Pusa Sandesh, *etc.*
- ☆ **Muskmelon:** Pusa Sarbati (Katana x PMR-6), Punjab Sunehari (Hara Madhu x Edisto), Hisar Madhur (Pusa Sarbati x 75-34), *etc.*
- ☆ **Watermelon:** Arka Manik (IIHR-21 x Crimson Sweet), Pusa Bedana 'triploid' (Tetra-2 x Pusa Rasal), *etc.*
- ☆ **Pumpkin**: Kashi Harit (NDPK-24 x PKM).

Backcross Breeding

Backcross method consists of crossing of F_1 with one of the parents (recurrent parent followed by selection of genotypes for specific characters. One to three backcrosses may be made till desired plant type is attained. This method is generally applied to transfer simple inherited characters, like resistance, stamina unimproved variety. Two types of parents are involved in this breeding method, one as sterility and desirable morphological traits to the parent (high yielding) and other as donor parent (low yielding but possessing specific desirable trait). This method is commonly used for development of resistant varieties. By using the Indian muskmelon as a resistant source, crosses were made with common type at University of California. Following backcross breeding, the resistant line to powdery mildew was selected and released in the name of Cantaloupe 45 (PMR-45). This resistance is governed by single dominant gene.

Heterosis Breeding

Heterosis for earliness, fruit size, fruit weight, flesh thickness and TSS (muskmelon and watermelon) have been reported in cucurbits. Heterosis for total yield ranges from 20-100 per cent in muskmelon, 20-60 per cent in watermelon, 35-40 per cent in cucumber and 30-60 per cent in bottle gourd, pumpkin, summer squash and bitter gourd. In India, the first commercial hybrids developed in vegetable were Pusa Manjari and Pusa Meghdoot in bottle gourd (1971) at IARI, New Delhi. Cucurbits are highly cross pollinated crops; therefore it is necessary to develop the inbred lines. If varieties/lines are maintained as pure line by inbreeding, these lines can be used as inbred lines. These inbreds will be crossed in different mating design (Diallel, Line x Tester) to test the specific and general combining abilities of the lines. Once a good heterotic parental combination is identified, the hybrid seed can be easily produced by adopting any one of the following method-

- Pinching of staminate flower buds in female parent before anthesis and hand pollination by male parent- All monoecious cucurbits.
- Pinching of staminate flower buds before anthesis in female parent and insect pollination- All monoecious cucurbits where isolation distance is available.
- Use of gynoecious line (female parent) and insect pollination from pollen of male parent under isolation- Cucumber and bitter gourd.
- Emasculation and hand/insect (at proper isolation) pollination- Muskmelon.
- Use of staminate sterility and insect pollination- Muskmelon.
- Chemical suppression of staminate flowers and insect pollination (at proper isolation).

Several heterotic combinations have been identified in different cucurbits for yield and yield contributing traits with quality attributes. Several F_1 hybrids in different cucurbits have been released after evaluation.

Mutation Breeding

Generally mutations are heterozygous and recessive. Therefore the mutant phenotypes are not expressed in M_1 generation. Several chemical mutagens (ethyl methane sulphonate, diethyl sulphate ethyl amine *etc.*) may be used for inducing the mutation. Besides, physical mutagens X-rays, gamma rays, neutrons, *etc.* also may be used for induction of mutation. In cucumber seed treatment of Cum-15 with 0.05 per cent ethylamine for 24 hours has given 7.6 per cent plant population resistant to *Meloidogyne* spp. In bitter gourd, gamma radiation and EMS (ethyl methane sulphonate) treatment led to an increase in genetic variability. Seed treatment of snake gourd by 18 kr X-rays has given superior mutant for fruit size, yield and high punicic acid content in seed. MDU-1 and PKM-1 varieties of bitter gourd and ridge gourd, respectively have been developed by mutation breeding.

Polyploidy Breeding

Alternation in chromosome during mitosis and meiosis can be induced by treating plant parts with colchicines is called polyploidy. This method of breeding has been commercialized in watermelon. The triploid F_1 hybrid has been developed by using tetraploid and diploid lines. Triploid plants produce seedless fruits after being pollinated by diploid pollen which stimulates parthenocarpy but ovule fails to develop because of the sterility accompanying the triploid condition. Seedless watermelon variety Pusa Bedana is developed at IARI, New Delhi using tetraploid line Tetra-2 and deploid line Pusa Rassal.

Sex Expression in Cucurbitaceous Crops

Cucurbits are polygamous crops. There is wide range of variation in sex forms. Each species is characterize by a specific sex expression but variation within also exist. The following sex forms have been also observed.

1. **Monoecious:** Staminate and pistillate flowers are found in same plants separately. This is most common sex forms in the majority of cucurbits like-ash gourd, bottle gourd, sponge gourd, ridge gourd, bitter gourd, cucumber, pumpkin, summer squash and snap melon.
2. **Andromonoecious:** Staminate and hermaphrodite are found in same plant separately. This is pre-dominant sex form of muskmelon and very rare in some watermelon. Ridge gourd, smooth gourd and cucumber in varying proportion.
3. **Dioecious:** Staminate and pistillate flowers are produced on separate plant. Pointed gourd, ivy gourd, sweet gourd and spine gourd are among the cultivated cucurbits that exhibit this sex forms.
4. **Gynomonoecious:** Pistillate and perfect flowers are produce separately on the same plant. This sex forms is reported in certain stalk of cucumber.
5. **Gynoecious:** Plants bear only pistillate flowers. For example, some line in cucumbers and bitter gourd.
6. **Androecious:** Only staminate flowers are produced. For example segregating population of cucumber, ridge gourd and muskmelon.
7. **Trimonoecious or Gynoandromonoecious:** Staminate, pistillate and hermaphrodite flowers are produced on the same plant in various proportions. This is highly unstable sex forms. Such form is mate within some breeding lines of watermelon, muskmelon, cucumber and ridge gourd.
8. **Hermaphrodite:** Staminate and pistillate organs are found on the same flower. It is found in 'Satputia' variety of ridge gourd.

Factors Affecting the Sex Expression

The appearance of staminate, pistillate, hermaphrodite and various sex forms largely depend upon environmental factors (Temperature and Photoperiod) and plant growth substances which modified the sex forms.

- ✰ High temperature and large photoperiod independently or in combination shift femaleness to maleness.
- ✰ Low temperature and short photoperiod shift maleness to femaleness.

- ☆ High nitrogen with adequate moisture and high atmospheric humidity produce more female flower.
- ☆ Exogenous application of plant growth regulator modified the sex ratio applied at two to four true leaf stages.
- ☆ Application of etherl @ 250 ppm which release the ethylene gas increase female tendency in cucumber, muskmelon, squash and sponge gourd.
- ☆ Application of NAA @ 100 ppm at two to four true leaf stage increase femaleness in sponge gourd.
- ☆ Maintenance of gynoecious line in cucumber by spray of GA_3 1500-2000 ppm or silver nitrate or silver thiosulphate.
- ☆ Application of TIBA (Tri-iodo-benzoic acid) @ 25-50 ppm increase femaleness in watermelon.
- ☆ Application of boron @ 3 ppm increase femaleness in bottle gourd.

Section-IV

Vegetable Seed Production Technology

Chapter 31

Vegetable Seed Industry

Introduction

Production of quality seed in our country has been inadequate and insufficient and so, seed were imported. Quality seed of vegetables were being imported from country like Australia, U.K., U.S.A, Germany, *etc.* up to 1939. During 18th and 19th century Portuguese and Britishers introduced temperate vegetable in India. In 1945 some private companies such as Sutton, Pocha, *etc.* began quality seed production of temperate vegetables in Quetta (now in Pakistan) and Kashmir valley. These companies formed an All India Seed Producers Association (AISPA) in 1946. After the partition of the country the Central Vegetable Breeding Station (CVBS) at Katrain (HP) was established by Govt. of India in 1949. Now it is known as IARI regional station, Katrain. Now large number of private sector seed companies, SAUs and ICAR Institutes have strengthened vegetable seed sector.

The Indian Seeds Act

In the year 1963 the Govt. of India set up National Seeds Corporation in the public sector to organize the development of a sound seed industry in India. The discipline of certification was maintained by NSC alone. In order to bring seed production under a legal umbrella the Indian Seed Act was enacted by parliament in the year 1966. The seed rules formed under the Seed Act were notified in 1968 and have been amended from time to time. This law came into force on October 2, 1969 in all the states and union territories of the Government of Indian. The objective of this act was to regulate the quality of seeds of all the crops being marketed for agricultural purposes. For this purpose labeling of seed bags was made compulsory for all types of seed sale related to notified varieties. Thus the two important features of this act were as follows:

- ✰ Compulsory labeling
- ✰ Voluntary certification

Broadly this act ensures as follows:

1. Different classes of seed must conform to the prescribed standards of purity and germination as detailed in the minimum seed standard of crop.
2. Seed should be marketed after proper tagging.
3. The tag should be providing details of purity of the seed along with complete name and address of producer.

The Indian Seed Act (1966) was amended in 1972, Seed (control) order was issued in December 1983, State and Central Seed Certification board has been established under section 8(a) of the Act amended in 1972.

Main Features of Indian Seed Act

1. Govt. of India to advise both central and state govt. on matters arising out of administration of this act.
2. Establishment of a central seed laboratory.
3. Establishment of a state seed laboratory.
4. A provision of notification of variety by Govt. of India.
5. The minimum limits of germination and purity of seed.
6. Establishment of state seed certification agencies.
7. Establishment of central seed certification board.
8. Appointment of seed analysts in state seed laboratory.
9. Appointment of seed inspector for determining, seed purity and related traits for sale.
10. Seed means seed of food crop, fodder, cotton, jute, tuber, bulb, rhizomes, root, cutting all vegetables, *etc.*

Govt. of India launched National Seed Project (NSP) in October 1976 with World Bank assistance. Since April 1980 ICAR took the continuance of project under Natinal Seeds Proect (NSP), ICAR is guiding, coordinating and promoting breeder seed production and seed technology research in selected ICAR institutes (4) and SAUs (about 60). State Seed Corporation (13) and Certification Agencies (19) have also been established to produce and make available quality seeds to growers in time and at affordable prices.

Seed Reforms

The department of agriculture and co-operation, ministry of agriculture and farmers welfare constituted a seed policy review group under the chairmanship of Dr. M.V. Rao (VC of APAV Hyderabad at that time) to examine the policy changes required in the seed sector in the context of the changing environment.

Principle recommendation of this group:

1. Compulsory registration of varieties to be introduced.
2. Amalgamation of seed control order with the provisions of Seed Act.
3. Self-certification by the seed producing agencies is provided.
4. Notification of seed testing laboratories.
5. Establishment of national seeds board in place of central seed committee and CSCB.
6. Increase in penalties for offences.
7. Review of labeling of certification standards.

New Seed Policy (1988)

A new seed policy on seed development was announced by the Govt. of India on 16 September 1988 which was implemented from October 1, 1989. The main objective this policy was to provide to the Indian farmers the best seed or planting material available anywhere in the world to increase farm productivity, income and export earnings. In order to give intensive to import of seed, the import duty on seed was reduced to 15 per cent from the earlier 100 per cent (90 to 105 per cent) and vegetable seed was kept under open general license.

Present Status

A separate NSP on vegetables was approved and initiated by IIVR (erstwhile PDVR) Varanasi in 1994. For breeder seed production ICAR had set up 4 units at ICAR institutes and 10 units in SAUs. In 11th Five Year Plans, the NSP (vegetables) has been merged with AICRP (VC) though the breeder seed production programme is continuing as such. India possesses varied agroclimatic zones, experienced farmers, viable seed industry and legislation, *etc.* for production of quality vegetable seed. Thus a scientifically proven and commercially viable road map encompassing the future strategies is running smoothly with involvement of private seed companies on a large scale. As a matter of fact, currently, about 90 per cent of vegetable seed production and sale is in the hands of private sector seed companies, both multinationals and Indian seed companies.

Strengths

1. A well developed and knitted seed multiplication and distribution system linked with several ICAR institute/SAUs/NS.
2. A network of 19 CSA and more that 90 notified STL to legally assure the quality of seed to market.
3. A large number of varieties available suited to varied agroclimatic condition. This makes the selection easier for taking up production in area.

4. A very fast development of private seed companies for supply of seed particularly those of hybrid seeds. About 90 per cent of vegetable seed market is in the hands of private seed companies and farmers are happy to have access to good quality seed including the imported ones by private seed companies.

Weaknesses

1. Vagaries of weather resulting in production of poor quality seeds.
2. Very low or no indents for new improved varieties due to ignorance about the performance of newly developed improved varieties.
3. Problems in lifting of produced seed against indents, *etc.*
4. Regulatory environment is a bit complex and tough and at times creates hurdles in the smooth conduct of vegetable seed business.
5. Legislation and decision making takes to long time. Seed bill drafted in 2004 is again under debate as seed bill 2019.

Future Strategies

1. Development of organized private sector.
2. Making rules and regulations to facilitate seed business with adequate IPR protection.
3. Popularization of public sector seeds, their production and distribution, may be by involving KVKs.
4. Financial support to public sector seed organizations.

Private Seed Companies

There are several PSC that take up the work of quality seed production.

- ☆ Plant Breeder Right/Plant Variety Right Act - **2001**.
- ☆ Seed Act – **24th December 1966.**
- ☆ Seed Rules – **1968.**
- ☆ Seed Amendment Acts - **9th September 1972.**
- ☆ Seed (Controls) order- **1983.**
- ☆ The Protection of Plant Varieties and Farmers Rights Act (PPV and FRA) – **30th October 2001.**
- ☆ The Protection of Plant Varieties and Farmers Rights Rules – **2003.**
- ☆ The Biological Diversity Act – **2002.**
- ☆ The Biological Diversity Rules – **15th April 2004.**
- ☆ International Seed Testing Association (ISTA) – **1924.**

Chapter 32

Seed Production Techniques

Introduction

Seed are product of ripened ovule after fertilization or reproduction by pollen and some grow within the mother plant. The embryo is developed from the zygote and seed coat from the integuments of ovule. Seeds have been more important developments in reproduction and success of gymnosperm and angiosperm plants.

Classes of Improved Seed

There are five classes of improved seed.

(1) Nucleus Seed

Nucleus seed is the initial seed of an improved variety which is always limited in quality. It is produced by the originating plant breeder.

Main Features

- ☆ **Production**: It is produced at the experimental farms of the concerned research institute or agricultural university under the supervision of original plant breeder or sponsored plant breeder.
- ☆ **Purity**: It is genetically and physically 100 percent pure.
- ☆ **Certification**: For nucleus seed certification is not required.
- ☆ **Use**: Nucleus seed is used for the production of breeder seed. It is not meant for general distribution. However, if it is more in quanity and remains unutilized, it can be used as breeder seed also.

(2) Breeder Seed

Breeder seed is the progeny of nucleus seed. It is produced under the strict supervision of original or sponsoring plant breeder at the research farm of the concerned crop research institute or agricultural university.

Main Features

- ☆ **Production**: Breeder seed is produced in isolation from other varieties. The isolation distance differs from species to species.
- ☆ **Purity**: Breeder seed is genetically 100 per cent pure and physically at least 98 per cent pure. The genetic purity is maintained by proper roguing.
- ☆ **Certification**: Certification is not required for breeder seed. The seed plot is monitored by a monitoring team which consists of original or sponsored plant breeder and one representative each from National Seed Corporation and State Seed Certification Agency, not below the rank of Deputy Director.
- ☆ **Use**: Breeder Seed is used for the production of foundation seed. It is not meant for general distribution.
- ☆ **Label colours and size**: Golden yellow and size 12 x 6 cm.

(3) Foundation Seed

Foundation seed is progeny of breeder seed.

Main Features

- ☆ **Production**: It is produced by national seed corporation under the strict supervision of research scientists and experts from NSC. Production of foundation seed is taken up at the seed multiplication farms of governments and research farms of ICAR institutes and also on cultivator's fields. Proper isolation distance is adopted for the production of foundation seed which varies from crop to crop.
- ☆ **Purity**: Foundation seed is genetically cent percent (99 per cent pure) however physical purity of 98 per cent is permissible. The remaining 2 per cent includes inert matter.
- ☆ **Certification**: In case of foundation seed, certification is required which is under taken by state seed certification agency.
- ☆ **Use**: Foundation seed is used for the production of certified seed. It is not meant for general distribution. Foundation seed stage I can be used for production of foundation seed stage II under certain emergent circumstances.
- ☆ **Label colours and size**: Foundation seed label colours is white and size is 15 x 7.5 cm.

(4) Certified Seed

Certified seed is the progeny of foundation seed foundation stage I or foundation stage II seed.

Main Features

- **Production**: It is produced on the field of progressive farmers under the strict supervision of state seed certification agency. This work is also taken up by the NSC, if required. Proper isolation distance is adopted which varies from species to species.
- **Purity**: It has genetic purity of 100 per cent (98 per cent) and physical purity also of 98 per cent.
- **Certification**: This class of seed requires certification by seed certification agency. For certification the seed must meet rigid requirement of purity and germination and other parameters as prescribed under field standards meant for each specific crop.
- **Use**: Certified seed and available for general distribution to the farmers for commercial crop production. Under special and emergent circumstances, certified seed can be permitted one more cycle of production and the produce is labeled as certified seed stage II beyond which no more production cycle is permitted.
- **Label colours and size**: Colours blue and size 15 x 7.5 cm.

Seed Multiplicationa and Release of a Variety

A. Multiplication of Seeds of a Variety

1. **Nucleus seed:** The seed is maintained by the particular breeder who evolved a particular variety. The nucleus seed will be 100 per cent genetically pure conforming to the varietal characters of a particular variety. The nucleus seed is utilized for raising the breeder seed.
2. **Breeder seed:** The breeder seed will be multiplied from the nucleus seed in the research stations by plant breeders. The breeder seed will be utilized for raising the foundation seed by the State Department of Agriculture. Every year the Director of Agricultural will place the indent of breeder seed to the Seed Division, Ministry of Agriculture and Farmers Welfare and further allocation for production to the state agricultural universities/ICAR institutes is decided in the all India coordinated crop research workshops. As per allotment, the university will take up breeder seed production in the research stations. The breeder seed plot will be monitored by the monitoring team to verify the varietal characters and genetic purity of that particular crop. The monitoring team members will include a Plant Breeder, Deputy Director of Agriculture (Seed certification) and a nominee from National Seeds Corporation. The monitoring team will visit the seed production plot twice in a crop growth period, *i.e.,* at the time of flowering and at the time of harvest.
3. **Foundation seed:** From breeder seed, the foundation seed will be raised in state seed farms. This foundation seed production plot is to be certified

by the seed certification agency. The foundation seed is utilized for raising certified seed production.

4. **Certified seed production:** Done either by the Agricultural Department or by individual farmers after paying a nominal registration fee. The seed production plot will be certified by the seed certification agency and after that the seed will be sold to farmers.

B. Steps Involved in Release of a Variety

After identification of the best cultures from the segregating generation or any other source it has to undergo the following trials.

1. **Row yield trial (RYT):** For every 10th row there will be a check entry and the trial will be non-replicated.
2. **Replicated row yield trails (RRYT):** From the row yield trial, the best cultures will be tested in RRYT along with appropriate check. The best entries from RRYT will be carried forward to preliminary yield trial.
3. **Preliminary yield trial (PYT):** Replicated trial conducted with appropriate checks. PYT will be conducted normally for two seasons. While conducting, PYT, the best entries will be nominated to all India trials also. Screening for biotic and abiotic stresses will be done during PYT stage. The best entry will be carried to comparative yield trial. The entries entered All India trial will be tested as per standard operating procedure in the crop project. Normally, in all India coordinated trials, there is three stage testing (Initial Evaluation Trial-IET, Advance Varietal Trial I-AVT I and Advance Varietal Trial II-AVT II) after which the variety is identified in the workshop by a committee constituted for this purpose.
4. **Variety release proposal:** The scientist in-charge of the culture will propose the culture for release as a variety. There is a proforma for variety release. This proforma will contain all the information about the culture, *viz.*, Parentage, parent's morphology, cultures morphology, key characters of the culture for identification, agronomic practices, pest and disease resistance, quality characters and yield trial results. The variety release proposal will be discussed by Director of Research and Scientists. After approval the proposal will be presented before Variety Release Committee.
5. **Variety release committee:** Central variety release committee now named as Central Sub-committee on Crop Standards, Notification and Release of Varieties for Agricultural Crops/Horticultural Crops is headed by DDG-Crop Science/DDG-Hort. respectively for agricultural and horticultural Crops and the proposals are submitted to Dy. Commissioner Seeds (Quality Control) in the MoA and FW, New Delhi. The constitution of committee is notified well in advance. After discussion, based on merit the CVRC will approve it for release. Then the culture will be released for

general cultivation. The State Variety Release Committee is headed by Director Agricuture/Director Horticulture with appropriate members.

6. **Notification of the variety:** For certified seed production, the variety is to be notified by the central variety release committee, Delhi. After release of the variety for notification purpose the information will be furnished in the prescribed proforma. At that time details about all India trial will also be furnished. After notification only, a variety can be multiplied under certified seed production.

Seed Certification

Seed certification is a legal system which ensures production of high quality seeds in terms of purity and germination the task of seed certification is undertaken by State Seed Certification Agency. This work is also taken up by NSC in those states where SSCA does not exist. There are four basic requirements for SC.

1. **Improved variety**: A genotype released for general cultivation either by the central variety release committee or by a state variety release committee is known as a improved variety. It is essential to select an improved and notified variety for seed certification.
2. **Genetic purity:** Genetic purity refers to absence of seed of other variety of the same crop as well as of other crops. Contamination by the seed of other crop is permitted from 0-0.5 per cent certified seed and 0-0.1 per cent foundation seed of different crops.
3. **Physical purity**: The freedom from inert matter and defective seed is known as physical purity. Inert matter includes non-living material like sand, pebbles, soil particles, straw, *etc.* The physical purity of 98 per cent is permissible in majority of crops.
4. **Germination**: The germination should be as per the standard fixed by Indian Seed Act 1966.

Minimum Isolation Distance for Crops

	Foundation Seed (m)	*Certified Seed (m)*
(A) Self-pollination		
Cowpea	50	25
Tomato	50	25
(B) Often cross-pollination		
Brinjal	200	100
Okra and chilli	400	200
(C) Cross-pollination		
Cucurbits	800	400
Cabbage	1600	1000
Cauliflower	1600	1000

	Foundation Seed (m)	Certified Seed (m)
Radish/Turnip	1600	1000
Onion	1000	400
Carrot	1000	800
Palak	1600	1000

Seed Producing Organizations

In India there are two types of seed producing organization, *viz.*, public sector and private sector. The public sector organization includes NSC, SSCs, SSCA but private sector includes various seed companies.

1. National Seeds Corporation

The NSC was established in March 1963 to take up the work of quality seed production and promote seed industry in the country. The NSC started functioning in July 1963. The head quarter of NSC is in New Delhi in the Pusa Campus. The responsibilities of NSC were more clearly defined with the inception of National Seed Project (NSP) in 1976.

Main Function

- ☆ **Production of breeder seed**: The requirement of breeder seed of various crops is assessed and arrangements are made for its production through crop research institute of ICAR and SAUs.
- ☆ **Production of foundation seed**: The production and distribution of the foundation seed of the varieties of national importance is the responsibility of the NSC.
- ☆ **Production of certified seed**: The production and marketing of the certified seed of vegetables is taken up.
- ☆ **Marketing**: Interstate marketing of the seeds of national varieties lies with the NSC.
- ☆ **Consultancy services**: The NSC provides consultancy services related to designing, procurement and installation of seed processing plant, equipments, *etc.*
- ☆ **Imparts training**: The NSC organizes short term training courses to the personnels engaged in seed production work.
- ☆ **Coordination**: NSC provides co-ordination in the production of certified seed by SSCs.
- ☆ **Established**: Takes up seed certification work in those states where SSCA has not been established. It is also taking up import and export work of seed.

2. State Seed Corporation

The National Seed Project was launched in 1976 to establish SSC from 1976 to 1987 SSC have been established in nine states namely Uttar Pradesh, Andhra Pradesh, Bihar, Kerala, Karnataka, Maharastra, Odisha, Punjab and Rajsthan. Their main functions are the distribution of certified seed in the respective state. In Uttar Pradesh, Tarai Development Corporation (TDC) was established in 1969 to produce certified seed which proved very much successful. Now, TDC has been included in Uttarakhand and is known as Uttarakhand State Seed Production Agency and UP has its own State Seed Production Agency headquarted at Lucknow.

3. State Seed Certification Agencies

SSCA are located in 18 different states including those states where SSC have been established. The main function of the SSCAs is certification of seed in the respective state.

Organizations Associated with Germplasm

There are two types of organization, *viz.*, international and national which are associated with germplasm.

1. International Plant Genetic Resources Institute (IPGRI) old IBPGR

IBPGR established in 1974 then change named in December 1993.

2. National Bureau of Plant Genetic Resources

It is established by ICAR in 1976 in New Delhi. In India plant introduction started in 1946 at IARI New Delhi in the division of Botany. In 1961 a separate division of plant introduction was established under leadership Dr H.B. Singh. NBPGR has several regional stations like (i) Shimla (HP), (ii) Jodhpur (RJ), (iii) Akola (MH), (iv) Kanyakumari (KR), and (v) Shilong (Meghalaya), *etc.*

3. Quarantine

There are chance of the entry of new disease, insects and weeds when the material collected and introduced from other countries. To prevent entry of new disease, insects, weeds an Act was formulated in India in 1914 which is referred to as Destructive Insects and Pest Act. In India quarantine work under these organization.

- ☆ NBPGR – **New Delhi**
- ☆ Botanical Survey of India – **Calcutta (Kolkata)**
- ☆ Directorate of Plant Protection, Qrantine, and Storage- **Faridabad (Haryana)**
- ☆ Directorate of Seed Research (DSR) - **Mau Nath Bhanjan (UP)**
- ☆ Central Seed Laboratory – **New Delhi**
- ☆ Central Seed Certification Board (CSCB) – **1972**
- ☆ Indo-American Hybrid Seed (IAHS) Bangalore – **1965**

- ☆ Protection of Plant Varieties and Farmers Rights Act. - **2001**
- ☆ Trade Related Aspects of IPR Agreement (TRIPS) - **1994**
- ☆ Indian Patent Act - **1970**
- ☆ Indian Patent Rules - **2005**
- ☆ The Union for the Protection of New Varieties of Plants (UPOV)- **1991**
- ☆ The Copyright Act - **1957**
- ☆ The Copyright Rules - **1958**
- ☆ Trade Marks Act - **1999**
- ☆ Trade Marks Rules - **1999**
- ☆ The Design Act - **2000**
- ☆ The Design Rules - **2001**
- ☆ The GI Act - **1999**
- ☆ The GI Rules - **2002**
- ☆ IC Layout Design Act - **2000**
- ☆ IC Layout Design Rules - **2001**
- ☆ PBR/PVR = Plant Breeder Right/Plant Variety Right Act - **1994**

Control Measures Adapted by Seed Certification Agency

1. **An administrative check on the origin of the propagated material:** Source seed verification is the first step in seed certification programme.
2. **Field inspection:** Regular field inspection of the seed plot is carried out to check varietal purity. Prescribed minimum isolation distance, physical admixture, presence of designated disease and objectionable weeds not beyond prescribed minimum standards.
3. **Sample inspection:** This is done in the lab to assess planting value of the seed.
4. **Bulk inspection:** There is provision under seed certification procedures to undertake bulk inspection to check the homogeneity of the bulk seed lot produced as compared with the standard sample drawn.
5. **Control plot testing:** The samples drown from the source seed and final seed produced are grown in field along with the standard samples of the variety under consideration.
6. **Grow out test:** The possibility to prove genuineness of cultivars by grow-out test is based on the hereditary characteristics of plants. It is easier to determine the cultivars purity in self fertilizing species than in cross pollinated species observation are mode during the whole growing period and deviation from the standard samples are recorded.

Phases of Seed Certification

Manly six board phases:

1. Verification of seed source, seed class and other requirements of the seed used for raising the seed crop.
2. Receipt and scrutiny of applications from the seed growers.
3. Inspection of seed crops in the field to verify its conformity to the prescribed field standards.

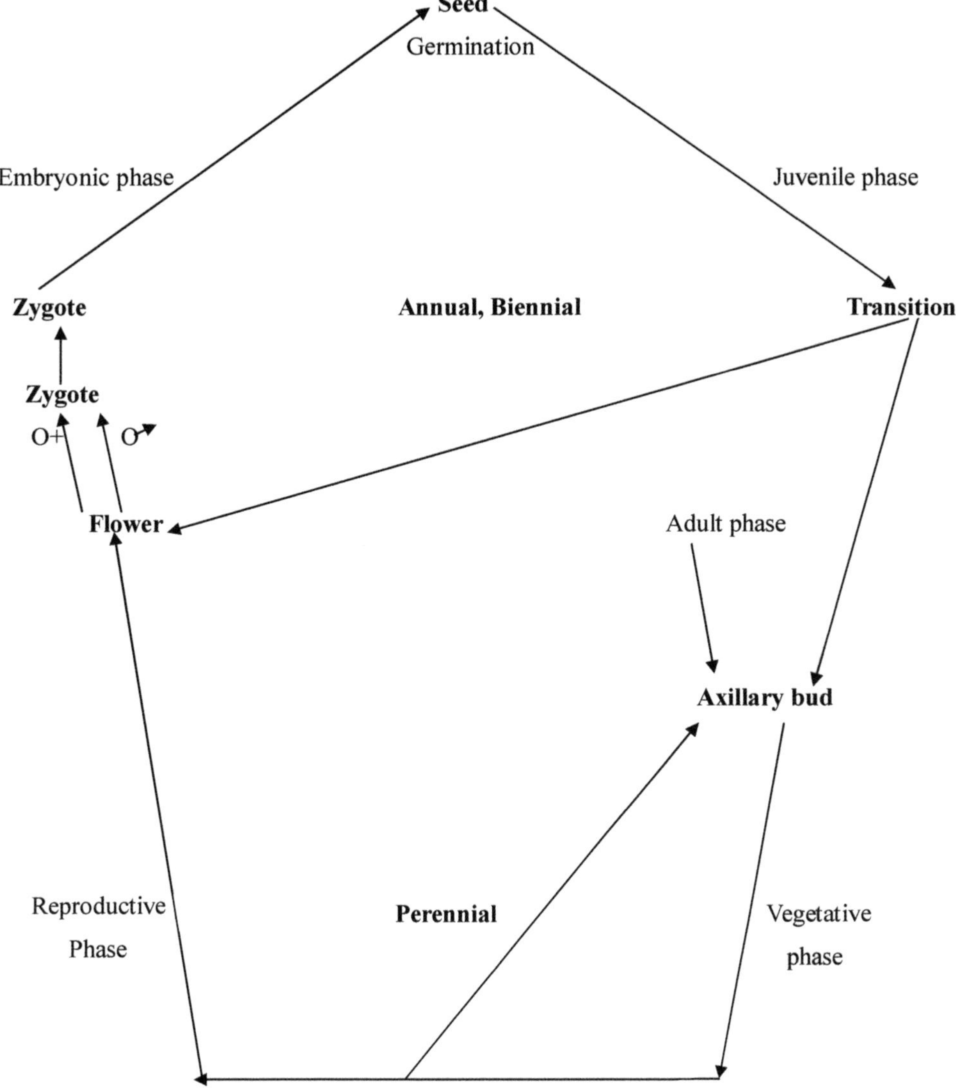

Seedling Cycle in Plants Showing different Phases.

4. Supervision and monitoring at post harvest stage including seed processing and packing at seed processing plant.
5. Drawing of samples and arranging for analysis of seed samples to variety conformity to the prescribed seed standards.
6. Grant of certificate issue of certification tags, labeling, sealing, *etc.*

Seed certification is a legally sanctioned system to variety quality control of vegetable seed production and multiplication at various stages which consists of different components *viz.*:

- A designed authority.
- Defining clearly vegetable seed germination.
- Notified vegetable variety/hybrid.
- Agronomic requirements.
- Recognized institute/vegetable breeder.
- Laboratory tests.
- Control plot testing.

Procedure of Seed Certification

1. Verification of seed source.
2. Field inspections to conform the prescribed land or field standards.
3. Proper supervision at pre and post-harvesting.
4. Seed sampling and testing in seed testing laboratories.
5. Tagging and sealing.

Main Step of Seed Certification

- Receipt and scrutiny of applications.
- Inspection of vegetable seed fields.
- Approval or rejecting the seed fields.
- Seed sampling and analysis.
- Tagging and sealing.
- Control plot testing.
- Revocation of certificate.

Seed Certification Agency

Under the National Seed Policy 20 State Seed Certification Agencies have been established. A Central Seed Certification Board has been set up at the centre to act as a self regulating agency. CSCB in the apex body to modify or fix new seed quality standards. The statutory bodies and various agencies involved in seed certification program under Seed Act- 1966 are as under.

1. **Central Seed Committee-** Under the supervision of union ministry of Agriculture Govt. of India.
2. **Central Seed Certification Board-** The Govt. of India has constituted the said board with the main aim to deal with all problems related to seed certification. Central Govt. provided financial assistance to the State Seed Certification on re commendation of the said board.
3. **State Seed Certification Agencies-** Recommend by central seed committee.
4. **Seed Testing Laboratories-** The quality attributes evaluated in seed testing laboratories are seed germination, moisture content, vigour, physical and genetic purity, free from seed borne disease and insect infestation. The standard seed methods for vegetable crops are approved by ISTA. The Central Seed Testing Laboratory is located at IARI New Delhi. Present time more than 180 seed testing laboratories in India.

Seed Quality

The term seed quality looks very simple and is in common use by the industry personnel and often there is seed quality control labs or seed quality assurance (QA) labs is most of the organizations involved in seed production, processing, packing and marketing. The Federal Seed Act dictates that seed quality is determined by two principal factors as:

Seed Purity

Seed purity tells the seeds man and consumer how much unwanted material is present in the desired pure seed and specifies the nature of each contaminant. Seed germination defined as the emergence and development from the seed embryo of those essential structures which for the kind of seed in question are indicative of the ability to produce a normal plant under favorable condition.

Seed Vigour

It is the sum total of those properties of seed which determines the potential level of activity and performance of the seed or seed lot during germination and seeding emergence. It is defined as per 1977 meeting of the International Seed Testing Association (ISTA) group held at Madrid (Spain).

Quality Control

It is reasonable assurance of the identity and quality of the planting material (vegetable seed/seedlings) a buyer is buying. The main purpose of quality control is to maintain and make available to the public (farmers) high quality seeds and propagating materials of superior varieties/hybrids so grown and distributed as to maintain high genetic quality, identify and purity. The important components of quality control of seed are:

1. Seed standards
2. Field inspections
3. Seed testing

Seed Testing

Seed testing is very important for distinguishing the good quality seed from the inferior one to ensure the good quality planting material. The International Seed Testing Association (ISTA) established in 1924 has developed uniform rules and regulations for seed testing. These rules prescribe testing techniques based on scientific evidence which are accurate within the stated statistical limits and practicable for every day operations. The trained manpower and a good laborator with standard equipment and are the pre-requistes for seed testing programme. The seed testing process can be sub-divided in the following heads:

- Sampling
- Moisture test
- Purity analysis
- Germination evaluation
- Biochemical test for seed viability
- Seed health testing

Sampling

A sample from a seed lot is obtained by taking small portion (primary sample) at random from different portion in the lot and combined together (composite sample) from this sample smaller samples (submitted samples) are taken for the laboratory. A part of submitted sample is known as working sample which is subjected for quality test in the laboratory.

Moisture Test

There is close association between seed moisture and seed longevity. There for moisture content of seed must be determined before storage or when it is offered for sale. For all crops seed moisture should not be at or greater than the prescribed limit given by Indian Minimum Seed Certificates Standards (IMSCS). The basic objective of moisture determination of seed in to prevent losses in seed viability till seed is planted for commercial or other purposes.

Method of Moisture Test

Three basic steps is follow:

1. Weighing the original seed sample.
2. Drying the sample for a specified period.
3. Weighing the sample after drying.

$$\text{Seed Moisture content per cent} = \frac{\text{Loss in seed wt after drying}}{\text{Final dry weight of seed}} \times 100$$

In general two methods are determining seed moisture:

1. **Air oven:** In which seed sample is dried in an over after taking precise weight. The temperature and duration of drying period vary from crop to crop, *e.g.,* cope crop, brinjal, chilli, onion and radish of 130°C/for 17 hr. and pea, okra, cowpea, tomato, cucurbits, carrot, 130°C/for 1 hr.
2. **Seed moistures meters:** In this metods use of seed moisture meter for measuring of moisture test in seeds.

Special Requirements

In some case pre drying is also necessary. This is required when seed grinding is essential and seed moisture is higher than 17 per cent.

Seed Purity

It is mainly purity test of two types:

(A) Genetic Purity

A genetically pure seed lot must have all the seed of the same genetic makeup, *i.e.,* of the variety in question. The accepted official method for testing the genetic purity of a seed lot is the-grow out test and some test by gel-electrophoresis, RFLP, AFLP, RAPD, microsatellites, *etc.*

(B) Physical Purity

It is purity by three methods:

- ☆ **Pure seed:** It is the proportion of seed lot that is represented by the species under investigation. It will include all the botanical varieties. The pure seed fraction does not exclude immature under sized, shriveled, diseased or germinated seed provided it is possible to identity the seed correctly.
- ☆ **Other seed:** It includes seed and seed-like structures of any plant species other than the concerned crop species. This is done by taking help of the distinguishing features set out for pure seeds of crops.
- ☆ **Inert matter:** It includes seed, seed-like structure and other matters as follows:
    - **Seed and seed like structure:** It can be pieces of broken or damaged seeds achenes, mericrarp, one-half of the original size or less and removed seed coat and structure define in pure seed in which it is readily apparent that not true seed is present.

- **Other matter:** It includes, soil, sand, stones, stems, leaves, pieces of bark flower, nematodes galls, fungus bodies and all other matter that is not seed.

Seed Germination Tests

The general procedure of seed germination test are follows:

- ✰ Sampling and preparing the replicates.
- ✰ Preparing the substratum.
- ✰ Using suitable equipment/space for germination.
- ✰ Meeting special requirement if any.
- ✰ Evaluation of seedlings.
- ✰ Calculating the results.

Biochemical Test for Seed Viability

Teterzolium Test (TZ Test)

TZ test was developed during 1940's by Prof. George Lakon of Germany. It is a quick easy and popular method of testing seed viability. This test utilizes the chemical 2, 3, 5 triphenyl tetrazolium chloride or bromide, which is a cream or light yellow colour compound. This chemical is reduced to an insoluble red coloured "Formazan" in the presence of dehydrogenase group of enzymes. These enzymes are essential for seed respiration, and hence also for seed viability. Seed viability is judged on the basis of the pattern and the intensity of red colour developed in their embryos. In some crop, *e.g., Brassica* spp. seed may be placed directly in the TZ solution. But seed coat of many species is impermeable to chemical so that a preconditioning may be required for adequate embryo-chemical contact. In general this is achieved by soaking the seeds in water at 30°C for few hours to allow complete rehydration of seed tissues in some cases are mechanism operation such as cutting or puncturing of the seed required for proper embryo chemical contact. Thus preconditioned seeds may be handled as follows:

- ✰ Seed are placed directly in TZ solution, *e.g.,* Pea and beans.
- ✰ Alternatively prior to placing the seed in TZ solution they may be bisected longitudinally, *e.g.,* Tomato.
- ✰ The seed may be subjected to pricking below the embryo or they may be bisected laterally, *e.g.,* In case very small seed of grasses.
- ✰ The seed coat may be removed, *e.g.,* Mellon and cotton.
- ✰ The seed coat may be scratched above the embryo, *e.g.,* Lettuce.
- ✰ A piece of embryo may be excised and stained, *e.g.,* Pines.

The commonly used values of these factors are pH 6-8, temperature 30-35°C, 0.1 per cent, 0.5 per cent, or 1.0 per cent TZ solution and treatment duration of 20-48 hr. Overstanding makes the interpretation of result difficult.

Evaluation of Seed Viability

After staining the seed are removed from TZ solution, washed with water and classified according to staining pattern. During evaluation the stained seed can be kept in water for up to 48 hr. to permit evaluation of larger seed samples. The staining pattern enables a seed scientist to judge the live and dead area of embryo. The problem may be overcame by treating the seed with a lactophenol solution (lactic acid 1: glycerin 2: water 1) for 10-30 min.

TZ Test for Seed Vigour

Although the importance of TZ test for measuring seed viability is well known. Its use as vigour test is still not well established.

Advantages of TZ Test

- Seed viability can be estimated using TZ in a much shorter time as compared to germination test. Therefore, this test is quite suitable for rapid evaluation of a large number of seed samples.
- This technique can be used even with dormant seeds or where seedling growth is extremely slow. Alternatively this test can be combined with germination test to detect the dormancy.
- Since TZ test require a close examination of seeds, seed analyst can often get an idea of the reasons for poor viability of seeds.
- Since dicot seed need not be damaged during TZ test the same seeds could be utilized for further investigation.

Disadvantage of TZ Test

- Evaluation of staining pattern and their interpretation in terms of seed viability is difficult requires considerable experience and is a pain staking work.
- The total work involved in the test requires a greater number of man hours than does the germination test.
- This test alone is unable to differentiate between dormant and non dormant seeds.
- It is not possible to detect the presence of harmful micro-organisms to germinating seed ling.
- Injury caused to seed by micro-organisms, fungicides and fumigants may not get detected.

Seed Health Testing

It refers to the presence or absence of disease causing organisms such as fungi, bacteria, virus, nematodes, deleterious, physiological conditions, *etc.*

- ☆ The inoculum carried through the seed may spread disease in the field.
- ☆ Seed lots from exotic sources may introduce disease.
- ☆ The information may supplement germination results.

The Method Seed Health Test

1. Examination without incubation:
    - ☆ Direct examination.
    - ☆ Examination of imbibed seeds.
    - ☆ Examination of organisms removed by washing.
2. Examination after incubation.
3. Examination of plants.
4. Other techniques.

Chapter 33

Development of Seeds

Introduction

The embryo is developed from the zygote and seed coat from the integuments of ovule. Seeds are important development in reproduction and success of gymnosperm and angiosperm plants.

Baccate

In botany a berry is a fleshy fruit within a stone produced from a single flower containing one ovary. A plant that bears berries is said to be bacciferous or baccate. A fruit that resembles a berry weather it actually is a berry or not can also called baccate.

1. **Angiosperms (enclosed seed):** These are produced in a hard or flesh structure called a fruit that encloses the seed for protection in order to secure healthy growth. Some fruit have layers of both hard and fleshy materials.
2. **Gymnosperms (naked seed):** In which no special structure develops to enclose the seeds which begin their development naked on the bracts of cones. However, the seed do become covered by cone scales as they develop in some species of conifers. If, ovaries are not formed and the ovule and hence the seeds are exposed. This is the basis for their nomenclature naked seeded plants. Two sperm cells transferred from the pollen do not develop the seed by double fertilization but one sperm nucleus unites with the egg nucleus and other sperm is not used. Sometimes each sperm fertilizes an egg cell and one zygote is then aborted or absorbed during early development.

Different between Angiosperm and Gymnosperm

Angiosperm	*Gymnosperm*
This group consist of flowering ornamental and all vegetable and edible fruits	It include all of the conifers, cedar redwood, junipers, cypress, pine, *etc.*
They are also the source of world hardwoods	Softwood such as pine fir are used to make paper, lumber and plywood
Angiosperms have ovule that are enclosed in a carpel	Gymnosperms do not produce flower
A carpel consists of stigma and style and an ovary	Gymnosperms rely on the airborne transfer of their pollen
The seed of angiosperm contains either one/two cotyledons a seed leaf	Most of gymnosperms lack vessels in the xylem (wood)
It is the flowering plant	It is the non-flowering plant

Parts of Seed

1. Cotyledons

The seed leaves attached to the embryonic axis there may be one (mono) or two (dicot). The cotyledons are also the source of nutrients in the non-endospermic cotyledons in which case they replace the endosperm and are thick and leathery. In endospermic seeds, cotyledons are thin and papery. Dicotyledons have the point of attachment opposite one another on the axis.

(A) Monocotyledons

These are the plants in which the single cotyledons is found and these are known as monocotyledonous.

- ☆ **Cotyledon-** Single cotyledons termed as scutellum in contact with endosperm through an epithelial layer.
- ☆ **Plumule-** Embryonic shoot covered by protective layer called coleoptiles.
- ☆ **Redicle:** Embryonic root covered by protective layer called coleorhizae.
- ☆ **Seed coat:** Fused with pericarp.
- ☆ **Endosperm:** Bulky part within which lies the small embryo, massive and starchy endosperm.
- ☆ **A leurone layer:** Special tissue surrounding the endosperm.

(B) Dicotyledons

Plant having more than one cotyledons are called as dicotyledonous.

- ☆ **Cotyledons:** Seed leaves and food storage structures provide nourishment to the developing radicle and plumule.
- ☆ **Plumule:** Embryonic shoot.
- ☆ **Radicle:** Embryonic root.

☆ **Seed coat:** Testa- Thick after layer.

Tegmen- Thin after layer.

☆ **Hilum:** Scar on seed coat through which the seed is attached to the fruit.

"Inside the seed coat is the embryo consisting of an embryonal axis and two cotyledons often fleshy and full of reserve food materials. At the two ends of embryonal axis are present the radicle and the plumule".

2. Embryo

In botany a seed plant embryo is part of seed, consisting of precursor tissues for the leaves, stem (hypocotyls) and roots (radicle) as well as one or more cotyledons. Once the embryo begins to germinate and grow out from the seed it is called as seedlings.

3. Ovule

In seed plants the ovule is the structure that gives rise to seed. It consists of three parts. The integument forming outer layers, the nucellus (or remmant of megasporengium) and the female gametophyte (formed from haploid megaspore) in its centre.

4. Epicotyls

The embryonic axis above the point of the attachment of cotyledons.

5. Plumule

The tip of the epicotyls has a feathery appearance due to the presence of young leaf primordia at the apex and will become the shoot upon germination.

6. Hypocotyle

The embryonic axis below the point attachment of cotyledons connecting the epicotyle and redicle being the stem root transition zone.

7. Redicle

The based tip of the hypocotyls around in to the primary root.

8. Nucellus

It is the central portion of an ovule in which the embryo sac develops.

9. Sperms

Pollen is a fine to coarse powdery substance comprising pollen grains which are male micro-gametophytes of seed plants which produce male gametes (sperm cells).

10. Funicle

The funicle or seed stalk which is attached the ovule to the placenta and hence ovary or fruit wall of pericarp.

11. Micropyle

It is small pore or opening in the apex of the integument of ovule where the pollen tube usually enters during the process of fertilization.

12. Chalaza

The base of ovule opposite the micropyle where the integument and nucellus are joined together.

Plants generally produce ovules of four shapes-

1. **Anatropous:** It is found curved shape.
2. **Orthotropous:** Straight with all parts of ovule lined up in a long row producing an uncurved shape.
3. **Campylotropous:** Ovule have a curved mega-gametophyte often giving the seed a tight "C" shape.
4. **Amphitropous:** Where ovule is partly inverted and turned back 90° on its stalk (the funiculus).

Types of Seed

The seed are of the three types for the seedlings:

1. Orthodox Seed

Seed are long lived and can be successfully dried to moisture content as low as 5 per cent without injury and are able to tolerate freezing. It is also termed as desiccation tolerant. In fact the life span of orthodox can be prolonged with low moisture content and freezing temperature. *Ex-situ* condition of orthodox seed is, therefore, not problematic. It is found in many annuals and biennials crops, *e.g., Citrus aurantifolia, C. annum, Hamelia patens, Lantana camera, Psidium guajava, Anacardium occidantale* and other horticultural crops.

2. Recalcitrant Seed

Seeds are remarkably short lived which cannot be dried to moisture content below 20-30 per cent without injury and are unable to tolerate freezing. Recalcitrant seeds are, therefore, also termed as desiccation sensitive seeds. Recalcitrant seeds are difficult to be successfully stored and their *ex-situ* conservation is problematic. It is because of their high moisture content that encourages microbial contamination and results in more rapid seed deterioration. Secondly storage of recalcitrant seed at freezing temperature causes the formation of ice-crystals which disrupt cell membranes and causes freezing injury. Therefore the plant that produces recalcitrant seed must be stored in growing phase (growing plant). Mostly tropical and temperate tree and shrubs plant and some common of plant that produce recalcitrant seed (which are generally larger than orthodox seed) include Avocado, Cacao, Coconut, Jackfruit, Litchi, Mango, Rubber, Tea, and medicinal plant.

3. Intermediate Seed

They show limited desiccation tolerance but are sensitive of freezing temperature, *e.g.,* Citrus, and Coffee seeds.

Chapter 34

Seed Production of Vegetables

Introduction

A seed is a mature ovule consisting of an embryo together with stored food, all enclosed by a protective coat. It is product of the ripened ovule of gymnosperm and angiosperm plants which results as a consequence of fertilization and some growth within the mother plant formation of seed completes the cycle of reproduction. The embryo develops from zygote and the seed coat from the integument of ovule. A seed has three main parts these are seed coat, endosperm and embryo. Embryo includes the root, the cotyledon and embryonic leaves. Seed is located inside fruit and a fruit is a mature ovary. In broad sense any plant part which is used for commercial multiplication of crop is called seed. Thus in case of potato stem and tuber are also known as seed. Improved seed results in (1) better germination (2) vigorous seeding growth (3) higher crop stand (4) better quality of produce (5) ultimately in higher crop yield.

Vegetable Crops

A seed is borne on fruit and fruit is a mature ovary and ovary is a typical part of flower. Flower in angiosperm, consists of four parts these are calyx/sepal, corolla/petals, androecium/stamens and gynocium/pistil, outer whorl of flower is collectively referred to as calyx. The individual components of calyx are called as sepal. Petals are collectively referred to as corolla. Petal is a unit of second inner whorl after sepals. Stamen is part of flower bearing the male reproductive parts composed of anther and filaments. Pistil is part of flower bearing the female reproductive parts composed of stigma style and ovary.

Seed Production Techniques in Vegetable Crops

Placentation

The manner of arrangement of seed within the ovary is called placentation.

There are three basic types:

1. **Parietal**: Seed are attached to the ovary wall.
2. **Axil**: The ovary is divided by partition called septa and seeds are attached to central axis.
3. **Free central (basal)**: Seed are attached to the central axis but there are no partitions or septa.

Basic Structure of Seed

Development of four megaspores through meiosis is known as mega-sporogensis. Three haploid megaspores normally degenerate. Development of female gametophyte or embryo sac from the functional megaspore through three successive mitotic divisions leads to one large cell enclosing eight haploid nuclei. This process is known as mega-gametogenesis, soon these nuclei arrange themselves within the enlarging embryo sac. Out of eight nuclei, three nuclei with cell wall at one end/pole are called antipodal cells. Two poler nuclei without cell wall occupy central place inside the embryo sac. There is egg apparatus consisting of one egg cell between two synergid cells. Two polar nuclei fuse subsequently to form one diploid (2n) nucleus. This arrangement leads to formation of seven celled structure known as mature gametophyte or embryo sac or megagametophyte.

During flowering pollen (haploid) is transferred from anther to stigma is called pollination. The mature pollen grain germinates on the stigma. A pollen tube with two male gametes grows down the style into the ovary until it reaches the embryo sac having 8 haploid nuclei within the ovule. Two male gametes from the pollen tube are discharged in to the embryo sac. One male gamete unites with the female gametes surrounded by two synergids to produce zygote. This process known as fertilization. The second male gamete fuses with two polar nuclei to produce endosperm which triploid (3n).

The relationship between floral parts and fruit or seed can be illustrated as follows:

- **Ovary (peicarp)** – Develops into fruit.
- **Ovule** – Develops into seed.
- **Integuments** – Develops into seed coat (testa).
- **Nucellus** – Develops into perisperm.
- **2 Polor nuclei + sperm nucleus** – Develops into endosperm.
- **Egg nucleus + sperm nucleus (zygote)** – Develops into embryo.

Global Seed Markets

According to estimates provided by International Seed Federation (ISF), global players in world seed production and marketing could be listed as follows:

- Bayer Crop Science - Germany
- BASF - Plant Science - Germany
- Dow Agro Science - USA
- Dupont -USA
- Limagrain - France
- Rijk Zwaan - Netherlands
- Syngenta - Switzerland

Indian Seed Market

The major players (Indian/joint venture/foreign seed companies) in Indian vegetable seed industry are represented by about 200 seed companies in India.

- Advanta Seed- **Bangalore**
- Tropical Seed - **Bangalore**
- Bayer Crop Science **(Seminis)**
- Bioseeds - **Hyderabad**
- Ganga Kaveri Seeds - **Hyderabad**
- Pioneer - **Hyderabad**
- Ankur Seed - **Nagpur**
- Bejo Sheetal - **Jalna**
- East-West Seed - **Aurangabad**
- Indo-American Hybrid Seeds - **Bangalore**
- Kaveri Seeds - **Sikandarabad**
- Mahyco - **Jalna Maharastra**
- Sygenta - **Mumbai**

General Principles of Vegetable Seed Production

Majority of vegetable crops are propagated by true seed while few like pointed gourd, coccinia, garlic, drumstick, *etc.* are propagated vegetatively. The production of vegetatively propagated crop is easy as compared to seed production of botanical (tree) seeds. The maintenance, preservation and multiplication of pure seed have greater significance as the variety is generated by the breeders after investing of very precious time, labour and money for that purpose. Therefore seed production of crop is done following definite procedures, which vary depending on the nature of the crop, *i.e.,* self or cross pollinated or asexually propagated and the type of variety (pureline, hybrid, synthetic or composite).

In general seed production involves the following steps:

1. **Selection of crop and variety:** Only those crop and varieties should be taken for the seed production which are high in demand, having potential or established market and performing well in the area.
2. **Selection of seed:** Unless initially the pure seed is procured from a reliable source, no effort can help to maintain its purity. The pure the seed, the lesser will be the effort involved in maintaining the varietal purity. It is therefore must to verify the source of seed.
3. **Reporting to monitoring/certification agency:** Improved seed production is subject to legal umbrella and it must undergo monitoring under the supervision of the authority defined in Seed Act. Therefore proper reporting by the seed grower or institution to the monitoring agency should be ensured.
4. **Selection and preparation of field**: Selection of the field is the key to success of a seed production programme. The field should be uniform in topography, fertility, texture, *etc.* which enables the seed crop to express all the passport traits uniformly. Soils are well fertile well drained, free from soil-borne disease and weeds. The proper tillage and herbicides should be used and a favourable seed bed should be prepared.
5. **Cultural practices:** This is vary with crop and geographic location. These practices also play a key role in obtaining good seed harvest likes adjusting the sowing time for seed crops, maintaining proper inter and intra row distance in such a way that plants express full vigour and their maturity may fall in the drier part of the season. Necessary steps should be taken for timely control of insect and disease. In case of severe disease, prophylactic spray is applied.
6. **Appropriate male: female ratio:** Male: female ratio is relevant particularly in case of hybrid seed production where the F_1 seed is the result of cross-pollination between two parents. In such cases only the female parent bears the hybrid seed. In order to minimize seed cost, male: female ratio is generally kept in favour of the female parent. The male: female ratio is decided on the basis of experimental investigations and may be 1:1, 1:2, 1:3, 1:4, *etc.*
7. **Maintenance of recommended isolation distance:** Isolation means keeping the seed plots away from other fields of the same crops and crossable species. Ensuring the proper isolation distance from the nearby field of the same crop and crossable species is the most important prerequisite for multiplication of seed vis-a-vis maintaining its purity.
8. **Field inspection:** Field inspection refers to the scrutiny of seed production plots by a team of qualified persons. The primary objective of field inspection is to ensure that seed production is being carried out of the designated variety and there is no contamination (physical

or genetically) beyond the specific maximum limits. It also ensure that the step necessary to minimize genetic and physical contaminations have been taken properly and in time field inspection is done by a team of qualified persons to verify the source of seed, prescribed land requirement male : female ratio and prescribed isolation or border row in case of hybrids, presence of off type and objectionable weeds, disease incidence of specified disease and similar other requirements. That may be specified for the crop concerned.

In general field inspection may be done at the following five stages:

- Preflowering
- Flowering
- Post flowering
- Preharvest
- Harvest stage.

In asexually propagated crop such as potato, sweet potato, pointed gourd, *etc.* growth stage like pre flowering and post flowering may not be relevant.

9. **Roguing:** Roguing is the removal of off type plants from a field. Off types are those plants that are phenotypically different from plant of the variety grown as seed crop. It is an essential component of seed production programme as it helps to attain desirable level of genetic purity. A rogue plant may cause genetic impurity due to mechanical mixture and by out crossing with the true to type plants of seed crops. Roguing is, however, continued till crop maturity in order to remove off type plants based on fruits that become distinguishable only at the later stages of development, *e.g.,* tomato fruit colour. In case of nucleus or breeder seed production roguing is performed under the direct supervision of the breeder or his associates.

10. **Harvesting:** Harvesting of seed crop at appropriate maturity is essential to avoid the wastage like dropping of overripe fruits and seeds in the field. In some cases such as okra and brinjal first picking for consumption or sale does not affect the ultimate seed yield but adds to the farmer's income in addition to other beneficial effects like reducing the disease inoculums build up. The seed produced on plant under moisture deficit are usually light and shriveled with poor vigour and are more prone to hard seededness. It is assumed that when soil fertility becomes limiting plant responds by producing less seed.

11. **Seed processing:** It is carried out to improve the quality of harvested seed through a series of operation, *viz.,* drying, cleaning grading, testing, seed treatments, bagging and labeling.

Basic Objective of Seed Processing

- Improving the seed quality by removing adulterants and unfit seed.
- Taking preventive measures that maintain seed viability.
- Provide information about seed quality standards to the perspective buyers.
- Making seed handling easier.

1. **Seed drying:** Seed should be dried to reduce the seed moisture to safer levels. Its basic purpose is to sustain the seed viability and longevity by minimizing attacks of storage pests and pathogens and avoiding heat damage due to higher respiration rates in moist seeds. Seed can be dried under natural condition (under sun) or in artificial condition using various types of driers. Artificial driers like bag driers, box driers, bin driers, flat storage drier, continuous flow tower drier, *etc.* can be used for drying seed to the desired level, *e.g.,* Moisture content (per cent) for drying crops: tomato, brinjal, chilli 8 per cent ordinary and 6 per cent vapour proof pack, okra 10 per cent ordinary and 8 per cent vapor proof pack.
2. **Cleaning and grading:** Cleaning is the separation of physical impurities like trash, dirt, weed seeds, *etc.* from the seed lots, whereas the grading refers to removal of undersized or underweight seed from the seed lot. Before cleaning and grading, preconditioning of seeds may also be done. Pre-conditioning also includes pre-cleaning, which refers to removal of large sized particles, light weight chaffy material and seed of small size from the lot. All cleaning operations basically involve separation of adulteration based on size, density or shape. Hand sieves can also be used for grading on small scales.
3. **Seed treatment:** It refers to exposing the seed to certain agents, physical or chemical which are able to protect them from pests and ensure good health of seeds and emerging plants. Its main aim is to protect the seeds from pests and pathogen such as bacteria, fungi, nematodes and insects. The most commonly used fungicides for seed treatment are captan, thiram and carbendazim. Equipment used for seed treatment is drum dry seed treater, slurry seed treater or mist-o-matic seed treater.
4. **Seed packaging:** Seed should be packed in smaller unit to avoid risks of physical gradients, particularly vapours pressures which arise in large bulks. Packaging in smaller units makes identification, transportation, handling and marketing easier. Packaging material and amount of seeds to be packed depends on several factors such as kinds of seeds, duration of storage, storage environment, moisture content, *etc.* Packaging materials can be classified into three types:

   (i) **Moisture vapours permeable**: Jute bag, cloth bag, paper bag.

(ii) **Moisture vapours resistant:** Jute bag laminated with thin polythene film.

(iii) **Moisture vapors proof:** Tin cans, polythene bag (700 gauges), aluminum foil pouch, *etc.*

5. **Seed storage:** The storage of seed by the growers is mainly for short term (3-9 months) but occasionally growers needs to store up to 18 months. The purpose of seed storage is to maintain the seed in good physical and physiological conditions from the time of harvest to the sowing.

Factor Affecting Seed Longevity in Storages

1. Kind of seed.
2. Initial seed quality.
3. Moisture content.
4. Temperature and relative humidity during storage.

Point to be Considered for Seed Storage

1. Storage in a cool and dry place.
2. Effective storage pest control.
3. Proper sanitation in seed store.
4. Drying seeds to safe moisture limits before storage.
5. Controlling storage conditions depending upon length of storage period and prevailing climatic conditions.

Pre-Storage Preventive Measures

1. Before arrival of new produce all processing and storage structure should be throughly cleaned and disinfected with residual sprays of insecticides, *e.g.,* Malathion 50EC (one part in 25 parts of water) @ 51/100 m^3 area (5 per cent dust @ 0.5g/kg seed).
2. The seed moisture content should be reduced to specified levels. Most of insects do not breed at such as low moisture content.

Chapter 35

Hybrid Seed Production

Introduction

In horticulture and ornamental or gardening hybrid seed is a seed produced by cross pollinated plants (pollination is the process by which pollen is transferred from the anther (male part) to the stigma (female part) of the plant thereby enabling fertilization and reproduction). This takes place in the angiosperms—the flower bearing plants. Hybrid seed production is predominant in the vegetable crop and home gardening. It is one of the main contributions to the dramatic rise in horticultural output.

F_1 Hybrid Seeds

An F_1 hybrid (or Filial 1 hybrid) is the first filial generation of offspring of distinctly different parental types. F_1 hybrids are used in genetics and selective breeding.

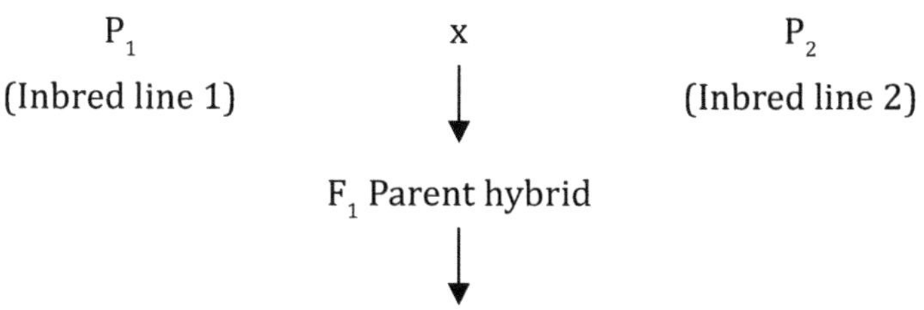

The first Indian vegetable hybrid was Pusa Meghdoot in bottle gourd which was released by IARI Pusa New Delhi in 1971. Two year later F_1 hybrids appeared in summer squash (Pusa Alankar) and cucumber (Pusa Sanyog) Hybrids result from the deliberate crossing of two different parent varieties from the same species. Pollen from male parent (pollen parent) will pollinate, fertilize and set seed in female parent (seed parent) to produce F_1 hybrid seed. F_1 refers to "first generation off spring" from these two distinct parent varieties. If plant seed saved from an F_1 hybrid variety then will not get the same result as the parent plant.

Share (per cent) of F_1 Hybrids

Cabbage 85 per cent, Watermelon 70 per cent, Okra 23 per cent, Tomato 80 per cent, Chilli 80 per cent, Cauliflower 50 per cent. Among the Asian countries, India is the 2nd largest production of hand pollinated hybrid seed in vegetables.

Hybrid Seed Production Male : Female Ratio

Okra- 1:4, Muskmelon- 1:3, Cucumber- 1:5, Watermelon- 1:6, Summer squash- 1:5, *etc.*

Commercial hybrid seed production demands crossing techniques which is easy and also economic to maintain parental lines. Only few crossing mechanism have been adopted for commercial hybrid seed production.

- ☆ Hand emasculation and pollination
- ☆ Self-incompatibility
- ☆ Decliny: monoecious and dioecious
- ☆ Male sterility

These techniques are specific to crop floral biology and flowering behaviour. These techniques have their own advantage and disadvantage. Based on the crop behaviour and crossing technique have been adopted for production of hybrid seeds commercially. Among different techniques self-incompatibility and sex expression have significance particularly in vegetable and flower hybrid seed production.

(1) Hand Emasculation and Pollination

Hybrid seed are produced manually by modifying the plant structure by removal of male organ from female plant before anthesis. This system is possible only when the male and female part of a single flowers or plants is separate. This is being adopted in bisexual or perfect flowers where the androecium is removed with case. The emasculation always done manually by the forceps in the late afternoon by removing the anther column or male part from female line the stigma of the emasculated flower is dusted with the pollen (pollen grain) of desired male parent next morning.

(2) Self-Incompatibility

Self-incompatibility is a mechanism which avoids self-fertilization through recognition of self-pollen in or on stigma on the female pistil. But when pollen from other plant carried by wind or insects is accepted and sets seeds.

Self-incompatibility will prevent self-pollination (inbreeding) and promotes cross-pollination (out breeding) and creates genetic variability.

Type of Self-Incompatibility

It is found mainly two types:

a. Heteromorphic
b. Homomorphic

(3) Decliny

Decliny or unisexuality is a condition in which the flower either staminate (male) or pistillate (female) occurs on the same plant or an different plants, *e.g.*, cucumber, watermelon, pumpkin, squash, asparagus, *etc.*

Types of Decliny

It is mainly of three types:

a. **Monoecious:** Unisexual flowers- where male and female flowers present on same plant separately, *e.g.*, Cucurbits, sweet corn, *etc.*
b. **Dioecious:** In this case male and female flowers are present on different plants, *e.g.*, Spinach, pointed gourd, asparagus, *etc.*
c. **Dichogamy:** The time of anther dehiscence and stigma receptivity are different forcing them for cross-pollination. The time gap between the two may vary from one day to many days. In protandry anther dehisce earlier than the stigma receptivity, in protogyny stigma become receptive earlier than the anther dehisce.

(4) Male Sterility

Hybrid seed production requires a female plant in which no viable male gametes are borne. Emasculation is done to make a plant devoid of pollen so that it is made female. Another simple way to establish a female line for hybrid seed production is to identify or create a line that is unable to produce viable pollen. This male sterile line is therefore unable to self-pollinate and seed formation is dependent and upon the pollen from the male line.

1. In hermaphrodite flower pollens are non-functional or inactive or sterile while, female gametes functions normally.
2. It is the inability of a plant to produce or to release functional pollen as a result of failure of formation or development of functional stamens, microspores or gametes.
3. Male sterility can be either genetic or cytoplasm or cytoplasm genetic. This prevents autogamy and permits allogamy.
4. Promotes heterozygosis, sterility is due to nuclear genes or cytoplasm gene or both.

5. In hybrid seed production process female is a male sterile line crossed with male fertility restorer line to get F_1 hybrid.

Hybrid Plants

Cross pollination of two compatible inbreds to produce a vigorous hybrid.

Constraints in Hybrid Seed Production

1. The hybrid seed are produced mostly by private seed companies and involve high production costs. High cost of hybrid seeds make ordinary farmer reluctant to use it as many of them cannot afford to purchase such costly seed. However, now the trend is changing and farmers are ready to pay higher price for good quality seed which can perform better, specially for hybrid seeds which have become major component in vegetable seeds in India.
2. Hybrid seed production is a very labour intensive enterprise particularly due to hand emasculation and pollination.
3. Hybrid seed production can be undertaken only by technically trained man power for the purpose.
4. Lack of popularization of public sector hybrids and unavailability of their seeds is sufficient quantities are another constraint in hybrid seed production.
5. Since the present-day ruling hybrids in vegetables are mostly from private sector which does not come under the national seed production chain thus there is no control on their prices. However, there is stiff competition in the market and hence prices canot be monopolized.

Indo American Hybrid Seed Company (IASH), Bangalore took the lead in the release of first tomato hybrid variety (Karnataka) and first capsicum hybrid variety (Bharat) in 1973 for commercial cultivation.

Seed Production of Tomato

1. **Nucleus seed**: It is produced by the concerned breeder or sponsored breeder. In nucleus seed crop, true to type single plants are identified and marked by putting tag on them. These are referred to single plant selection (SPS). Around 20 SPS are sufficient to obtain seed for raising one hectare breeder seed crop. These are planted at isolation distance of more than 50 m.
2. **Breeder seed:** The nucleus seed is sown to raise the breeder seed crop at an isolation distance of more than 50 m supported by regular inspection and roguing of off types from the breeder seed crop.
3. **Foundation and certified seed:** The foundation seed crop is raised from the breeder seed or foundation seed stage I and certified seed crop

is raised from foundation seed or certified seed stage-1.Tomato is self pollinated crop but cross-pollination to the extent of 2-5 per cent may occur. Therefore isolation distance of 50 m and 25 m are provided. At the time of final inspection off types in the field should not exceed 0.1 per cent in foundation seed and 0.2 per cent in certified seed.

Seed Extraction of Tomato

1. **Fermentation methods**: Red ripe fruits should be picked and crushed in non-metallic pots where profuse foam formation indicates that fermentation is completed. Fermentation of seed material should not be prolonged beyond 72 hr to get seed of high germination. Preferably the seed should be washed free of pulp after 48 hr.
2. **Acid treatment methods**: In case of urgency seed may be washed by using 1.0 to 1.5 liters of 35 to 38 per cent concentrated HCl/kg tomato fruit pulp. Fermentation is completed in about 30 minutes, clean seeds are obtained by stirring and washing procedure as quickly as possible.
3. **Alkali treatment**: Preparation 10 per cent sodium bicarbonate is mixed in equal volume of pulp. Fermentation completed over night.
4. **Mechanical seed extraction**: It is used for large scale operations put the ripe fruits into the mechanical seed extractor for crushing and separation of the seed and gel from the pulp, by acid 0.8 per cent solution of acetic acid for 24hr. at or below 21°C (1kg seed/10 liter of acetic acid).

Seed drying: Seeds are dried in the sun in case temperature exceeds 40°C. The seed drying under partial shade is desirable. Seed may be dried in seed dryer at moisture level 6-8 per cent.

Seed yield: Average seed yield are also found varies 80-100 kg/ha

Seed treatment to remove viruses: By 30 minutes in 10 per cent solution house hold bleach (0.525 per cent sodium hypochlorite NaCl) and 5hr in 5 per cent HCl.

Seed standard and seed tagging: In which foundation seed and certified seed are minimum germination 70 per cent and minimum purity 98 per cent without any admixture of other crop seed.

Index

N

O

P

Q

R

www.ingramcontent.com/pod-product-compliance
Ingram Content Group UK Ltd.
Pitfield, Milton Keynes, MK11 3LW, UK
UKHW021448280726
14060UKWH00001BA/298